A Practical Guide to Drug Development in Academia

Daria Mochly-Rosen • Kevin Grimes
Editors

A Practical Guide to Drug Development in Academia

The SPARK Approach

Second Edition

 Springer

Editors
Daria Mochly-Rosen
Department of Chemical and
Systems Biology
Stanford University School of Medicine
Stanford, CA, US

Kevin Grimes
Department of Chemical and
Systems Biology
Stanford University School of Medicine
Stanford, CA, US

ISBN 978-3-031-34726-9   ISBN 978-3-031-34724-5   (eBook)
https://doi.org/10.1007/978-3-031-34724-5

Editorial Contact: Charlotte Nunes

This Springer imprint is published by the registered company Springer Nature Switzerland AG
The registered company address is: Gewerbestrasse 11, 6330 Cham, Switzerland

Paper in this product is recyclable.

We dedicate this book to our SPARKees— Stanford faculty, postgraduate fellows, and students who have stepped beyond the confines of traditional academia to embrace the challenges of drug discovery and development. We are inspired by your efforts to translate your discoveries into new treatments that will benefit patients, and we have learned a great deal from each of you.

Preface

Our intention in writing and updating this volume, *A Practical Guide to Drug Development in Academia: The SPARK Approach*, was not to generate a comprehensive "how to" book; this topic cannot be taught in 282 pages. Rather, this volume addresses requests from our own SPARKees as well as from other academic institutions that have established or are planning to establish their own translational research programs. This book provides only a few pages on each topic that are part of the long process of drug development. The book is intended for novices who embark on this amazing path of translating basic research or clinical findings into new therapeutics or diagnostics to benefit patients. Ideally, you will read this book before you begin your translational effort, so that you will understand the path ahead—starting with the end in mind; planning what the final product will be ("target product profile" in industry lingo); and then creating a path back to where the project is today—doing it the industry way, with our own academic twist. This book should also be useful for students and postdoctoral fellows who plan to begin their career in the biopharmaceutical industry or who hope that their academic career will be translational in nature.

SPARK at Stanford is now in its 17th year. The program evolved slowly. We have found a certain formula that works well for our academic inventors and we hope that it will work for you as well. This book was written mainly by our SPARK mentors— the true engine behind the success of the program. It is not a complete guide to what needs to be done, but provides a general overview of the important topics in the development process.

Drug development is an applied science with a concrete goal in mind. It is also highly challenging, and academia can greatly help in this effort; not only through our inventions, but also by rethinking the drug and diagnostic development process, so that success rates will be higher and approval of the interventions will be faster.

Successful implementation of a translational research program requires securing experts in multiple disciplines (chemistry, biology, pharmacology, toxicology, medicine, regulatory science, statistics, and many more). Therefore, in addition to covering some key issues in each topic, we peppered the chapters with lessons that surprised us along our journey through biotechnology entrepreneurship. With these, we hope to demonstrate how drug development is not necessarily intuitive. You will find these in boxed text throughout the chapters.

After working on my own translational research projects and mentoring over six dozen other academic projects, I can summarize the following key lessons that I have learned.

1. Check your ego at the door—drug discovery is not about any one person; it is a true team effort, requiring experts in multiple disciplines. The weakest link is standing between you and a total failure.
2. Consult, consult, consult—there is always someone who knows much more than you about what is needed (preferably more than one person). Find these individuals and get their advice.
3. Always continue to apply your own judgment, as even experienced advisors can be wrong.

As the editors of this volume, Kevin Grimes and I hope that you will find it useful in your endeavor.

And now—let's SPARK.

Stanford, CA, US   Daria Mochly-Rosen
                                                                                          Kevin Grimes

Acknowledgements

We wish to express our sincere gratitude to Dr. Adriana A. Garcia for her tireless efforts to bring this volume to completion and Mary Romeo for careful editing of the text. Thanks also to our many contributing authors, who have generously shared their expertise and knowledge to help advance the cause of academic drug discovery. Special thanks to Dr. Phil Pizzo, the previous Dean of Stanford University School of Medicine, who endorsed the SPARK program from the start, and to our current Dean, Dr. Lloyd Minor, for his continued and unwavering support. Lastly, we owe a tremendous debt of gratitude to our SPARK advisors, whose generous and loyal support has been instrumental in SPARK's success. —*DM-R & KVG*

Contents

Getting Started

1

Daria Mochly-Rosen, Kevin Grimes, Robert Lum,
and Rebecca Begley

Daria Mochly-Rosen, Ed, Kevin Grimes, Ed.

In recent decades, we in academia have focused on advancing scientific understanding through basic research and have counted on the biopharmaceutical industry to translate promising discoveries into new therapeutics. Given recent developments, however, this paradigm needs to change. Pharmaceutical companies have drastically cut research budgets and basic research staff to decrease costs and improve short-term profits. As a result, we expect that fewer novel drug programs will originate in the biopharmaceutical sector.

Academic inventors can and should step in to fill this gap in the discovery pipeline. However, we often lack the expertise and resources to advance our projects through the applied science stage of drug discovery and development. This chapter introduces the process of drug development and highlights some important first steps: understanding the clinical needs, developing a target product profile (which defines the new drug's essential characteristics), and adopting a project management approach. These essential steps not only increase the likelihood of success but can also help decrease both the cost and time required to accomplish the goal. Translating

D. Mochly-Rosen (✉) · K. Grimes
Chemical and Systems Biology, Stanford University School of Medicine, Stanford, CA, US
e-mail: sparkmed@stanford.edu; kgrimes@stanford.edu

R. Lum
Concentric Analgesics, Inc., San Francisco, CA, US

R. Begley
SPARK at Stanford Advisor, Stanford, CA, US

D. Mochly-Rosen, K. Grimes (eds.), *A Practical Guide to Drug Development in Academia*, https://doi.org/10.1007/978-3-031-34724-5_1

discoveries from bench to bedside is a challenging but incredibly rewarding process, allowing us to advance scientific discovery and ensure that our government- and taxpayer-funded research translates into improved health for our society.

At SPARK, we know that our model works. SPARK is designed to assist academics in filling this discovery pipeline gap. Since the inception of the SPARK program at Stanford in 2006, we have worked on 242 projects, 200 of which have graduated from the program. These have yielded 51 startup companies, 28 technologies licensed to existing companies, and 21 noncommercial clinical trials. In addition, we have trained over 400 graduate students and postdoctoral fellows, who will apply their knowledge of drug discovery and development to their future careers in academia and industry.

Our SPARK model has spread globally into 23 countries on six continents, helping to build local pharmaceutical ecosystems. For example, SPARK Taiwan has cultivated over 300 teams, from which over 50 projects have led to a startup company, 30 have been licensed to an existing company, nearly 80 have entered clinical trials, and over 50 have received additional government funding for product development. SPARK Japan has supported 23 projects, of which 2 are in clinical trials and 3 have been licensed to a large company. SPARK Japan's teams have raised $7.9 M in public funding and $30 M in private funding. SPARK Berlin has accepted 56 projects into the program; 30 have completed the program, of which 12 projects have founded a startup or received follow-on translational funding. From SPARK Finland, 20 startups have been established out of 35 projects that completed the program. SPARK Colorado has accepted 42 projects into the program, 21 of which have successfully graduated, including 12 new startups. From the $4.2 M invested, SPARK Colorado has seen a return of $20.1 M to SPARK teams.

SPARK at Stanford has also been recognized in publications by others including financial expert Andrew Lo at MIT (Kim et al., *Drug Discovery Today* 2017) (Kim et al. 2017); Etzkowitz et al., *Manage Decis Econ* 2018 (Etzkowitz et al. 2020); and Gehr and Garner, *Cell* 2016 (Gehr and Garner 2016). SPARK was recently awarded the 2020 National Xconomy Award for Ecosystem Development, the Cures Within Reach 2020 Patient Impact Award, and the California Life Sciences 2022 Pantheon Award for Excellence in Academia, Nonprofits, and Research. Finally, the economic impact of SPARK at Stanford through licensing projects to startups and existing pharmaceutical companies was recently estimated at almost $2.9 billion.

We hope that you will join us in helping translate academic discoveries into solutions that benefit patients and society.

1.1 Advancing New Treatments from Academia to Biopharma

Daria Mochly-Rosen

In 2000, the Mochly-Rosen laboratory demonstrated that a rationally designed inhibitor of the delta isozyme of protein kinase C (δPKC) reduced infarct size when delivered right after heart attack by 70% in rat and pig models. The basic research that led to this result began more than a decade earlier. Our lab had studied protein–protein interactions and their specific role in PKC-mediated signal transduction.

Because we needed tools to probe these protein–protein interactions, we developed a methodology to design selective peptide inhibitors and activators of individual members of the PKC family of enzymes (isozymes). After confirming the effect of these modulators in vitro, we replicated these effects in cultured cells.

How did our lab begin studying heart attack? In 1997, using the peptide regulators of protein–protein interactions, we found that two PKC isozymes activated by adrenaline in the heart caused opposite effects: One increased and the other decreased the rate of contraction of cardiac muscle cells in culture. I presented our data at the American Heart Association Meeting, thinking that regulating the contraction rate of the heart would interest cardiologists. To my dismay, the report triggered no response from the audience.

Dr. Joel Karliner, then Chief of Cardiology at University of California San Francisco, was kind enough to point out the problem. As I was leaving the lecture hall, he told me that new approaches to regulate the rate of cardiac contraction were of limited clinical importance because there were good existing therapies. He advised, "Focus instead on determining the role of PKC in heart attack." He also suggested bringing a cardiology fellow into my laboratory "just to keep us informed." Mary Gray, MD, joined my group as we set out to examine the potential clinical uses for our basic research tools. In less than 3 years, my student, Dr. Leon Chen, and two cardiologists had shown that we could substantially reduce the infarct size of heart attacks in vivo by treating animals with the δPKC inhibitor. Surely, the pharmaceutical industry would now be pounding on our doors!

With the help of the Stanford Office of Technology Licensing (OTL), we secured intellectual property through patent filings. To our surprise, pharmaceutical companies were not remotely interested in our findings. We heard an assortment of reasons: "Rats are easy to cure," "peptides are not drugs," and "kinases are poor drug targets." Out of complete frustration that a potentially life-saving treatment garnered no interest from the industry, I founded KAI Pharmaceuticals with Dr. Chen. We visited scores of venture groups over 18 months attempting to raise funding for KAI. Finally, after a successful pre-Investigational New Drug application (pre-IND) meeting with the US Food and Drug Administration (FDA) in 2003 with our clinical advisor, Dr. Kevin Grimes, and a handful of other consultants, we convinced investors to fund us. In 2004, I took a leave of absence from Stanford to serve as chief scientific officer of the company for its first year. (More about the fate of KAI and lessons I learned from my association with the company are provided throughout the book.)

1.1.1 SPARKing Translational Research in Academia

While at KAI, it became clear that there must be many clinically valuable discoveries at Stanford's OTL that were not licensed. These inventions may be considered less attractive for many reasons including: (1) lack of proof-of-concept data in animals, (2) poorly characterized new chemical entity or drug with which the industry has no or little experience (e.g., our peptide inhibitors), (3) addresses a clinical indication

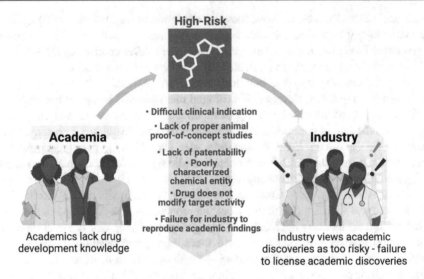

Fig. 1.1 Failure to license discoveries from academia. Academics lack drug development knowledge, which often results in drug candidates that are underdeveloped. As a result, industry perceives candidate development as too risky and most academic inventions fail to be licensed. (Created with BioRender.com)

known to be difficult (e.g., expensive clinical trials like Alzheimer's disease and/or indications where pharma had prior failures, such as stroke), or (4) involves a therapeutic target (e.g., a particular receptor) without a known drug that modifies its activity. Even something as promising as an entirely new therapeutic platform is often considered unattractive because of the long development time and uncharted regulatory path.

In other words, academic inventions are generally deemed to be premature and therefore too risky for pharma and/or investors (Fig. 1.1). Therefore, academic institutions must develop these discoveries further if they are to attract commercial interest. Some discoveries can be advanced directly to the clinic without commercial support, particularly when developing diagnostics or "repurposed" generic drugs. It is our social responsibility to step into this gap so that our discoveries will benefit patients.

1.1.2   What Is SPARK?

SPARK is a hands-on training program in translational research that I founded in 2006 based on my experience at KAI and co-direct with Dr. Kevin Grimes. SPARK's main mission is to accelerate the transition of basic discoveries in biomedical science to FDA-approved drugs and diagnostics. SPARK provides training opportunities in translational research to faculty members, postdoctoral fellows, and students.

Our goal is to move 10–12 new discoveries each year from the lab to the clinic and/or to commercial drug and diagnostic development.

Each autumn, we select approximately a dozen new projects to participate in SPARK. The program solicits short proposals from across the university to participate in a two-year cycle of training. SPARK selection criteria are quite simple:

1. The invention addresses an important unmet clinical need.
2. The approach is novel.
3. Two years of SPARK support will increase the likelihood that the invention will enter clinical trials and/or be licensed.

Finalists are invited to present their proposal to a selection committee of representatives from Stanford and the local biotechnology community. When preparing their presentation, the inventors follow a SPARK template that requires information beyond outlining the scientific background. After a brief introduction of the scientific rationale, presenters focus on the clinical benefits and basic requirements of their product (Target Product Profile, discussed in Sect. 1.4) and propose a development plan with specific funding requests, timeline, and milestones. In other words, the inventor is asked to plan from the final product back to the experiments that will generate it. We encourage this project management mindset—starting with the end in mind, a thinking process that is more prevalent among industry scientists than academic researchers—because novel discoveries can only advance toward a clinical therapeutic by following a disciplined path of applied science during development.

Importantly, unlike regular seed grants that go directly to the lab's account, SPARK funding (averaging ~$50,000/year for 2 years) is managed centrally and pays only for research to support preagreed upon milestones, money that has not been used in time reverts to the general fund pool and may end up supporting another SPARK project.

Dozens of experts from local biotechnology and pharmaceutical companies join our inventors at the SPARK meeting each Wednesday evening. SPARK's success depends on these experts who volunteer their time and agree not only to maintain complete confidentiality, but also to assign any inventions resulting from their SPARK advising to the university (Fig. 1.2).

1.1.3 I Love Wednesdays

We often hear the above comment as people reluctantly leave the room at the end of our SPARK sessions. Between 80 and 100 people meet every Wednesday evening throughout the year in a room bustling with energy and excitement. During the two-hour meeting, SPARK inventors (aka SPARKees) either present progress reports on their project or listen to an interactive lecture by an industry expert on a topic related to drug or diagnostic discovery and development. (This book briefly introduces the topics of these lectures.) Each inventor presents a progress report approximately

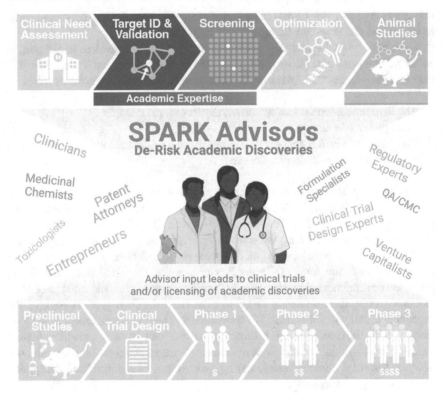

Fig. 1.2 The SPARK model. Typically, academics are well-versed only in the earliest stages of drug discovery (blue). SPARK aims to improve the percentage of academic discoveries that enter clinical studies (noncommercial) and/or are licensed by industry. SPARK educates participants on the multiple disciplines essential to successful drug development and provides project-specific mentorship by SPARK advisors (experts in all facets of drug development) throughout all stages of academic drug development. Abbreviations: *CMC* Chemistry, Manufacturing and Controls, *QA* Quality Assurance. (Partly created with BioRender.com)

every 3 months, and the amount of progress often surprises even the most experienced SPARK advisors. The benefit of the meetings comes from the strong commitment of the advisors to share their knowledge and experience in real time. This feedback is invaluable in helping SPARKees overcome challenges and find a path forward toward achieving their goals.

1.1.4 SPARK Track Record of Success

SPARK at Stanford is in its 17th year of existence. The educational value of the program is substantial for graduate students, postdoctoral fellows, and faculty. The experience is particularly helpful for trainees who are seeking positions in industry. However, a more quantitative measure of SPARK's success can be assessed by four parameters:

1. Licensing of projects (to a funded company).
2. Entry into clinical trials.
3. Publications.
4. Additional research funding awarded to SPARKees that they attributed to their work in SPARK.

A total of 200 projects have completed their participation in SPARK. Of these, 28 were licensed to existing companies, 21 were moved into the clinic (noncommercial), and 51 formed startup companies; a success rate of 50%! Note that 9 of the startups were acquired by established pharmaceutical companies and 3 startup companies went public. The financial analysis is equally impressive. When adjusted for inflation to current values, projects completing SPARK received $31 million in pilot funding while participating in the program. Following program completion, they have raised $2,772 million in commercial funding and an additional $103 million in external grants, a combined total that is 93 times the amount of funding that SPARK provided. As SPARK establishes a history of valuable projects, we see more industry interest in collaborating or licensing, even for early discovery-stage projects.

What this analysis lacks is the economic impact of new product sales on the academic institution. This is not an accidental omission; the impact can be measured only many years later. It takes ~12 years for commercialization of a new drug. If academic institutions invest in translational research hoping for revenue, there is a risk that their programs will become risk-averse, focusing on low-hanging fruit with limited impact on patient care or on indications that have large markets (e.g., another erectile dysfunction drug) rather than a true novel therapeutic for an unmet clinical need. If SPARK is rewarded for innovation, for getting programs to the clinic regardless of the commercial value, and for impact on the drug and diagnostic development process in general, together we may be able to have a substantial impact on improving patients' health and reducing healthcare costs.

1.1.5 Should Academia Be Engaged in Advancing Early Inventions?

You might believe the answer is obvious. However, I have learned over the years that not all of our faculty colleagues are supportive of our translational efforts and interactions with the life sciences industry. Here are some arguments I find compelling.

It is our social responsibility: Most of our research is supported by public funds. Our fellow citizens expect better health in return. Therefore, we should make an effort to ensure that our work can be translated to novel therapeutics or diagnostics, which improve outcomes and lower costs in healthcare. This will invariably require enlisting commercial partners.

It fits our education mission: Most of our graduates who do not choose an academic career will work in industry. It is therefore an opportunity to educate them on the development process and, through them, to educate the industry on what academia can contribute.

It is pure fun: Academics often hold the opinion that industry's work is applied science and therefore less intellectually demanding or gratifying. My year in industry and the years that followed taught me that this is incorrect. Drug development and diagnostic development are intellectually challenging and are exciting and worthwhile activities.

It is an opportunity: The success rate of drug discovery and development is still dismal, and the consequences of failures greatly impact our healthcare system. There is a special role and advantage for academics in improving public health through drug and diagnostic development. First is the cultural difference between industry and academia. While industry by nature is risk averse, academia gives higher rewards to risk takers—innovation and impact on the field are key components in faculty promotion and awards. In addition, academics are not burdened by knowledge of previous failures in industry; there is little published work on the topic, so we are free to apply new ideas to old problems. Further, academia can rely on the enthusiasm and brilliance of students who are the major engines of our research and innovation. Finally, there is a disincentive in industry to share information. On the other hand, in academia, all we learn is passed on through teaching and publications and thus can positively impact industry, which can translate into better health and lower healthcare costs.

> **Box 1.1: What Surprised an Academic?**
> Good science is important for raising funds from venture capitalists. Equally important are a clear, logical and realistic plan to develop a product, a strong team to run the company, and a positive attitude. We can't allow our egos to stand in the way. The rejection rates for trying to raise funds for your startup or license it to an existing company are even higher than those for paper submissions or grant applications. Your stamina will be tested as will your ability to learn quickly from each failed pitch. —*DM-R*

> **The Bottom Line**
> **SPARK has three missions:**
> 1. To accelerate the transition of basic discoveries in biomedical science to FDA-approved drugs and diagnostics
> 2. To educate the future workforce in translational research and better prepare them for jobs that they may hold in industry
> 3. To help improve the drug discovery and development process so that great academic ideas will translate to excellent solutions for patients
>
> The success of the program is largely due to the advice and mentoring of over 100 industry experts who volunteer their time and efforts, some for 15 years.

Recommendations
- Including a clinician in a basic research team can provide a real advantage when considering translational opportunities. When a basic discovery is made, the team has an opportunity to consider whether it also has clinical relevance. Understanding a clinical need is not necessarily intuitive; why not engage clinicians early to help identify how our discoveries may be put to clinical use?
- Translational research is complementary to basic research, but it should not be conducted at the expense of basic research. Medical schools will weaken themselves if the pendulum swings too far in favor of translational research.

Without a doubt, basic research is essential for our mission and should remain the primary focus of academic research. However, I strongly believe it is our responsibility as academic institutions to contribute to the development of drugs and diagnostics to benefit society.

1.2   Overview of Drug Discovery and Development

Kevin Grimes

Drug discovery and development is not for the faint of heart. The bar is high for a new molecule to receive regulatory approval for widespread clinical use—and appropriately so. As patients, we want our drugs to be both safe and effective.

The failure rate in drug development is quite high. Only 14% of drugs entering clinical study receive regulatory approval, and the failure rate is even higher during the preclinical phase of development. Given the complex array of drug-like behaviors that the new molecule must exhibit and the large number of interdependent tasks that must be successfully accomplished during development, this high rate of attrition is not unexpected.

The development cost for each successful drug is staggering, ranging from several hundred million to several billion dollars. The latter figure typically includes the cost of failed programs and the cost of capital. While an exceptionally well-executed program may be completed within 7 years, the norm is closer to 12 years and often much longer. Since patent protection for a new compound is granted for 20 years, the period of exclusive marketing after regulatory approval is typically in the range of 7–8 years, leaving a relatively short time for recovery of costs and generation of profit.

1.2.1   The Shifting Landscape

We are currently in a time of transition in the biopharmaceutical sector. The number of pharmaceutical companies has contracted through mergers and acquisitions as larger companies seek to fill their pipelines. Many profitable drugs have faced their "patent cliff"—increased competition from generic manufacturers entering the market after patent expiration. The "blockbuster" business model, which favored development of drugs for very large markets (statins, antihypertensives, drugs for type 2 diabetes, etc.), has fallen into disfavor as advances in "omics" allow for more tailored patient therapies. Previously ignored orphan diseases (diseases that affect <200,000 patients in the US) have become more attractive for several reasons: (1) Regulatory incentives effectively guarantee a period of marketing exclusivity, (2) clinical development costs can be substantially lower, and (3) "designer drugs" can command a premium in pricing.

Large biopharmaceutical companies have drastically cut their basic research staff because of pressure to decrease costs to improve profits. The amount of biotechnology venture funding for new biopharmaceutical startup companies is cyclical—reflecting the performance of the overall market and the biopharma sector. For example, the recent overheated initial public offering (IPO) market has resulted in lower returns that may temper investor enthusiasm for future IPOs. With less IPO competition, large pharmaceutical companies will pay less of a premium to acquire venture funded companies. In turn, venture capital companies will be more selective in which companies they fund and how much they invest. Changes in government policies might also impact investment in biopharma. For example, allowing CMS (Centers for Medicare and Medicaid Services) to negotiate drug prices will reduce premium prices for novel drugs and reduce investor enthusiasm for funding new companies.

If industry reduces spending on research and development, academics are well positioned to step into the breach, especially if there is institutional support for translational activities. This support can come through funding, the creation of core service centers (e.g., high-throughput screening centers, medicinal chemistry units, animal imaging centers, phase 1 units), educational programs in translational research and drug development, and a culture that values bringing new treatments to patients. By advancing our promising basic research discoveries toward novel therapies for unmet clinical needs, we academics will maintain our social contract with our fellow citizens who pay for our research and hope for better health in return. In addition, we contribute to our economy when successful academic programs enter the commercial sector as either startup companies or new programs at existing bio-pharmaceutical companies.

1.2.2   The Critical Path

Prior to obtaining market approval for a new drug, a number of complex steps must be successfully navigated (Fig. 1.3). While many of the steps must be accomplished

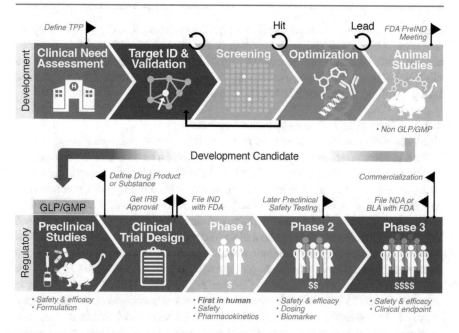

Fig. 1.3 General drug development pipeline. Abbreviations: *BLA* Biologics License Application, *FDA* Food and Drug Administration, *GLP* Good Laboratory Practice, *GMP* Good Manufacturing Practice, *IRB* Institutional Review Board, *IND* Investigational New Drug Application, *NDA* New Drug Application, *TPP* Target Product Profile

sequentially (e.g., demonstration of safety in animals before study in humans), drug development is very much an iterative process where a given step may be informed by or contingent upon many others. The following paragraphs introduce some of the critical steps in the development process. The timing of the handoff from academia to industry depends upon several factors, including cost of development and commercial attractiveness. For example, a repurposed drug may be fully developed within academia, whereas a costlier monoclonal antibody program may necessitate an earlier handoff to fund ongoing development.

1.2.2.1 Identifying the Opportunity/Target

Academics have discovered promising new drugs through a variety of approaches. Serendipity has played a role, as when Alexander Fleming combined critical observation with scientific acumen to discover penicillin. Unexpected side effects in early clinical studies have also been used to therapeutic advantage. This is how the anti-angina/antihypertensive drug sildenafil was repurposed as the first drug in the lucrative erectile dysfunction market.

Typically, however, academic drug discovery has been biology-driven—the result of hard work at the research bench. Novel associations are uncovered between specific proteins (or protein mutations) or pathways and one or more underlying diseases. The causality must then be proven through target validation studies using gene knockout/knock-in models, siRNA (small interfering RNA) gene silencing, or tool molecules that modulate the activity of the protein of interest. More recently, artificial intelligence (AI) applied to "omics," cell signaling pathways, or even clinical databases has played a role in target identification. These AI results are then brought back to the research bench for further validation.

1.2.2.2 Selecting the Therapeutic Approach

Once confidence in your target is established, it is time to consider the therapeutic approach. Most often, intracellular targets will require a small molecule therapeutic, whereas cell surface targets (e.g., CD20) and circulating bioactive molecules (e.g., tumor necrosis factor) may be amenable to monoclonal antibody (mAb) approaches as well. Some advantages of mAb therapeutics include a more predictable safety profile, less frequent dosing, and premium pricing. Disadvantages include the need for parenteral (intravenous or subcutaneous) dosing and higher early development costs, a major drawback for academic researchers. Some diseases are best addressed by replacing deficient hormones (e.g., thyroxine for hypothyroidism) or bioactive proteins (e.g., glucocerebrosidase in Gaucher's disease or erythropoietin for anemia associated with kidney failure). Recent advances have allowed novel therapeutic approaches for genetic diseases, including antisense oligonucleotide therapies, gene therapies, and genomically modified cell therapies. Future advances may include RNA editing, administration of stabilized mRNA (messenger RNA), and administration of exosomes or their miRNA (microRNA) contents.

1.2.2.3 Assessing Clinical Need

Before embarking on an expensive and time-consuming development plan, it is important to ensure the therapy will provide a clinical benefit. Take an unbiased look at clinical need, the suitability of the approach, and the feasibility of clinical development. This is best accomplished by a comprehensive review of the relevant literature and by extensive discussions with clinical experts and disease advocacy groups. The goal is to develop an outline of the clinical development plan, including route of administration, dosing regimen, efficacy endpoints, and duration of the trials. Obtaining help from a clinical trial design expert and a regulatory consultant can ensure that the team is on the right path. This subject will be discussed in greater detail later in this chapter.

Once you have determined your clinical indication and therapeutic approach, it is imperative that your team establish a Target Product Profile (TPP) . This critical document defines the essential characteristics of the final drug product and will serve as an important guide throughout the development process.

Box 1.2: What Surprised an Academic?
We as academics often assume that drug development, unlike our own basic research, is a rather mundane and straightforward process. In fact, those of us who have spent time in the biopharmaceutical industry have found that drug discovery and development lies at the intersection of basic research and applied science and requires a great deal of creativity and rigor. Drug development can be every bit as challenging and require even more persistence than traditional academic research. —*DM-R*

1.2.2.4 Determining the Preclinical Model

Unfortunately, rodents are much easier to cure than humans. A critical step in predicting the efficacy of the drug under development is to select an appropriate animal model for the clinical indication. While animal models are only imperfect approximations of the clinical disease, some models are more analogous than others. We should recognize that the administration of a neurotoxin to produce acute Parkinsonism in rodents has little similarity to chronic Parkinson's disease, which progresses over many years in patients. But aside from transgenic models for rare forms of familial Parkinson's disease or that overexpress alpha-synuclein, we currently do not have better models. Similarly, we might predict that occlusion of the middle cerebral artery in a rat would closely approximate a stroke in humans caused by acute occlusion of the same vessel. However, multiple new drugs that showed efficacy in rodent stroke models failed to show benefit in humans.

There are many reasons for this lack of predictive value. The animals are typically inbred in preclinical in vivo studies. The study animals are relatively young and frequently of one sex. They are fed the same food and follow the same sleep-wake cycles. Human subjects have diverse genetic backgrounds and come from a wider range of ages (often older) of both genders. Furthermore, human patients eat a varied diet, have diverse microbiomes, follow varied lifestyles, and may be taking several concomitant medications that could interfere with the new drug's absorption, metabolism, mechanism of action, or apparent treatment effect. Complementing animal models with studies in mixed cell cultures or three dimensional organoids derived from human induced pluripotent stem (hIPS) cells with relevant mutations can be equally important in validating a therapeutic approach.

1.2.2.5 Defining the Drug Candidate

In the case of small molecules, identifying a drug candidate typically requires designing a chemical or cell-based assay to identify activators or inhibitors of a target and then optimizing the assay for use in a high-throughput screening (HTS) facility. Most HTS centers have libraries containing between 10^5 and 10^6 compounds. Generally, a successful HTS will identify a few families of related molecules that have activity against the target of interest at low micromolar concentrations. An experienced medicinal chemist can help exclude certain hits as false positives that either interfere with the assay's reporter or exhibit exceptional promiscuity in targets. Those compounds that appear to be true hits can then be validated in a secondary screen. The most promising of these will become the lead molecule.

Once the team is satisfied that the lead compound truly modulates the target, a medicinal chemist can suggest chemical modifications to help optimize the desired molecular features, including potency (ideally activity in low nanomolar concentrations), selectivity for the desired target, solubility, bioavailability, duration of action, protein binding, plasma half-life, distribution to the target tissues, etc. This structure–activity relationship (SAR) analysis is an iterative process involving new chemical modifications and biologic testing to identify the most promising compound. The goal is to identify a drug that meets the prespecified criteria for efficacy, safety, and pharmacokinetics/ADME (absorption, distribution, metabolism, and excretion)—ideally with a wide therapeutic window (the ratio of toxic dose to efficacious dose). This process should culminate in designating a development candidate that satisfies the predefined advancement criteria in the TPP.

The final drug product will include not only this active pharmaceutical ingredient (API), but also excipients to help maintain stability (shelf life), control dissolution rate, and otherwise optimize the performance of the drug. Certain salts of the API may provide better solubility characteristics than others. Once the optimized formulation intended to progress into clinical studies has been identified, we can proceed with the more expensive IND-enabling preclinical studies.

1.2.2.6 IND-Enabling Preclinical Studies

Once the drug product is finalized, it is time to design and execute a series of rigorous preclinical studies to characterize the safety, pharmacokinetics, ADME, and interactions with drugs that will be given concurrently in the clinic. These studies must be carried out under Good Laboratory Practice (GLP) and are typically conducted at a GLP contract research organization (CRO). GLP entails a good deal of quality control and documentation to ensure that the studies are carried out in exactly the manner as stated. FDA provides guidance documents on its website regarding these studies, which must be completed before filing an Investigational New Drug application (IND) to begin a human clinical study. Because of the scope of work and documentation required, IND-enabling studies may cost several million dollars.

Prior to embarking on these expensive studies, it is prudent to arrange for a pre-IND meeting with FDA. The goal is to obtain general concurrence on the development plan and to ask specific questions regarding the drug product, proposed clinical studies, and preclinical development plan. Since the animal toxicology studies must predict safety for the human studies, they must be similar in route of administration, dosing regimen, and duration. Therefore, seek assurance that the proposed series of preclinical studies will be acceptable to FDA.

During GLP toxicology studies, animals must be dosed for at least as long as the intended clinical studies, so animal studies can only be designed after formalizing the clinical study design. The drug product used in preclinical studies must also be prepared according to GLP standards. Ideally, this GLP drug should be less pure than the clinical grade drug product that will eventually be dosed in patients. If the contrary were true, animal toxicology studies would not adequately reflect safety for patients, because the increased impurities in the clinical drug will not have been tested in animals. More expensive GLP reproductive toxicology, carcinogenicity, and long-term stability studies can often be deferred until before initiation of phase 3 clinical studies.

1.2.2.7 Obtaining GMP Drug Product

Drug product that will be used in the clinic must be manufactured and quality-tested according to Good Manufacturing Practice (GMP). GMP manufacturing requires exacting procedures and documentation and must be carried out at an experienced and certified facility. In addition to the manufacturing procedures, strict quality testing is performed at set intervals (e.g., every 3 months) under a variety of conditions to ensure that the drug is of highest quality. The drug product is tested to ensure the API has not degraded and that new impurities have not appeared. Parenteral formulations are also tested for sterility and for presence of endotoxins. The GMP manufacturing process and quality testing are resource intensive and quite expensive, so it makes sense to obtain several quotes and enlist a Chemistry, Manufacturing, and Controls (CMC) expert to evaluate facilities under consideration. FDA will often audit GMP facilities to ensure compliance.

1.2.2.8 Filing the IND

The IND contains three major sections. The clinical section contains the clinical protocol for the phase 1 clinical trial as well as the investigator's brochure, which describes the drug in detail and reports possible safety issues based upon preclinical animal safety studies. The preclinical section reports results of the GLP studies and any additional information that may be relevant to safety. The CMC section contains information regarding the API, formulation, manufacturing process, and quality control studies. Once the IND has been submitted, FDA has 30 days to respond with concerns, or clinical studies in humans may commence.

1.2.2.9 Clinical Development

Phase 1 studies are first-in-human studies primarily conducted to characterize the drug product's pharmacokinetics and determine its safety in people. Phase 1 studies are typically conducted in a small group of healthy volunteers. Occasionally, they are carried out in patients who stand to possibly benefit if the drug carries significant risk of adverse effects (e.g., neutropenia) or must be administered in an invasive manner (e.g., intracoronary or intraventricular). For example, cancer patients are often the subjects of phase 1 studies of chemotherapeutic agents since these drugs typically produce serious side effects.

Phase 2 studies are performed to explore the effective dose range or dosing regimen and to demonstrate efficacy. Often, the primary endpoint in phase 2 studies is a surrogate biomarker associated with disease progression rather than a clinical endpoint, since the latter would require a much larger study to reach statistical significance. For example, when studying a new drug for chronic heart failure, the study might be powered to demonstrate a difference in ejection fraction on serial echocardiograms rather than a change in the composite of hospitalizations and death. Once adequate efficacy is demonstrated for surrogate endpoints and the best dose(s) determined, the drug is ready for pivotal phase 3 studies.

Phase 3 studies are larger studies comprising a greater number of participants that are powered to clinical endpoints acceptable to FDA. Typically, two separate studies with an efficacy p-value of <0.05 are required for final drug registration. If the drug addresses a serious unmet need, FDA might allow a single study with a lower p-value (e.g., <0.01). Assuming phase 3 studies demonstrate both safety and efficacy, it is now time to compile the data into a New Drug Application (NDA) or Biologics License Application (BLA) and submit to FDA. Review of this final submission may take up to 18 months. If the project has Fast Track designation for a drug that addresses a serious unmet need, the review is expected to be completed in 10 months. FDA may request that an Advisory Committee comprised of external experts make a recommendation regarding final market approval, although FDA may concur or disagree with the Advisory Committee's recommendation. Once an approval is granted, the drug can be marketed in compliance with FDA regulations.

> **The Bottom Line**
> Drug discovery and development is a complex process involving many inter-dependent disciplines. Success requires creativity, persistence, some degree of luck, and a willingness to enlist the aid of experts in various fields.

1.3 Assessing Clinical Need

Kevin Grimes

As academic drug developers, we hope to translate our ideas into effective new therapies that will save lives, improve health and quality of life, and/or lower the costs of healthcare. We arrive at our therapeutic approaches in different ways. We may be basic research scientists who have discovered a promising new cellular target or pathway that plays a critical role in one or more serious diseases of which we have only a superficial clinical knowledge. We may be physician scientists or basic researchers who have dedicated our career to finding a cure for a specific disease with which we are intimately familiar. In either case, we need to call upon the collective wisdom of our peers, disease experts, and experts in drug development to ensure that our therapeutic approach will address the unmet needs of the patients in an optimal manner, and the unmet needs are great.

Despite impressive advances in drug therapy over the past 50 years, tremendous numbers of patients are in desperate need of effective new therapies for a wide variety of medical conditions. The list of diseases with inadequate or no treatments is daunting. Consider the following examples: pediatric diseases such as bullous skin diseases, inborn errors of metabolism, and autism spectrum disorders; obstetric disorders including premature birth and preeclampsia; global health challenges such as multidrug-resistant tuberculosis, chronic Chagas' disease, and newly emerging viral diseases; autoimmune conditions including progressive systemic sclerosis (scleroderma), systemic lupus erythematosus, and Crohn's disease; neuro-degenerative conditions such as amyotrophic lateral sclerosis, Huntington's disease, and Alzheimer's disease; and a wide variety of intractable malignancies. These examples are just the tip of the iceberg.

1.3.1 Starting with the End in Mind

Before we embark on a lengthy and costly campaign to develop a new drug, it is imperative that we understand why patients and providers will use our proposed product. What clinical problem are we solving? What specific unmet medical need will our product address? How will patients or the healthcare system be better off once our new drug is available? Are there known or predictable risks involved with modulating our drug's molecular target, and if so, is the risk-to-benefit ratio acceptable to our intended patient population? Will the drug delivery and dosing approach be acceptable to patients and providers? Will payers (insurers, health plans, Medicare, Medicaid) agree to pay for the new therapy?

1.3.2 Understanding Clinical Need

The first step is to understand the unmet clinical need. There are numerous reasons why a new therapeutic might be needed for a given condition. The following categories provide a framework for analyzing the necessity for a new drug for a clinical indication.

1.3.2.1 No Therapies Currently Available

Clinical need is most apparent when there are no effective treatments for a serious disease. Amyotrophic lateral sclerosis and advanced pancreatic adenocarcinoma are clear examples where current drug therapy has little to offer except palliation.

1.3.2.2 Need to Reverse or Arrest the Disease Process

For other serious diseases, we have therapies that reduce symptoms temporarily and even prolong life, but do not arrest disease progression. For example, current drugs for Parkinson's disease improve neurologic symptoms and improve quality of life, but do not prevent the relentless downhill course of the disease. Similarly, current therapies for idiopathic pulmonary arterial hypertension are vasodilators that do not arrest progression of the underlying pathology. Although symptomatic therapies are available for such diseases, there is a tremendous need for novel drugs that will modify the progression of the disease.

1.3.2.3 Severe/Unacceptable Side Effects

For many other diseases, current treatments may be effective, but cause serious or unwanted side effects. A few illustrative examples follow: (1) Hodgkin lymphoma was once a fatal disease but can now be cured in the majority of cases using a combination of chemotherapy and radiation therapy. Despite this success, patients frequently develop delayed but life-threatening cardiac toxicity from doxorubicin, one of the first-line chemotherapy drugs. (2) Corticosteroids can be life-saving treatments for a wide variety of autoimmune, allergic, or inflammatory diseases, but cause a litany of very harmful side effects. (3) The calcineurin inhibitors cyclosporine and tacrolimus, important components of immunosuppressive regimens following organ transplantation, can cause nephrotoxicity. Unfortunately, these drugs often damage the transplanted kidneys that they are protecting from the host immune system.

Many other very commonly prescribed medications cause unwanted side effects that affect the patient's health, quality of life, and even willingness to adhere to the drug regimen. Selective serotonin reuptake inhibitor (SSRI) and serotonin–norepinephrine reuptake inhibitor (SNRI) antidepressants commonly cause sedation, weight gain or loss, and sexual dysfunction. Metoclopramide, the first-line proki-netic drug for diabetic gastroparesis (delayed gastric emptying), can cause move-ment disorders including irreversible tardive dyskinesia. Clearly, patients would benefit tremendously from effective drugs that lack such undesirable side effects.

1.3.2.4 Patient Preference/Convenience/Cost

Generally, oral drugs that require less frequent dosing are preferable and improve patient adherence. Physicians rarely prescribe oral erythromycin (dosed four times daily for 7–10 days) now that FDA has approved azithromycin (dosed once daily for 5 days). Some drugs must be administered intravenously at an infusion center, which is inconvenient and costly. Alternative treatments that a patient can dose at home would be preferable.

Many new therapies, especially biological drugs, are prohibitively expensive. Less costly drugs would be a terrific boon to patients, insurers, and the healthcare system. New platforms for biological drug discovery, development, and manufac-turing could increase the success, shorten the timelines, and lower the costs of new therapies.

1.3.3 Suitability of Approach

After studying the unmet clinical need, the second step is to determine whether our planned therapeutic approach will provide an acceptable solution. For example, a peptide therapeutic injected subcutaneously twice a day might be readily acceptable for treating cancer but is a nonstarter for male pattern baldness. Speaking with phy-sician experts, patient advocacy groups, patients, and eventually FDA (and other regulatory agencies) will help us identify acceptable and ideal drug characteristics.

In the case of a serious or life-threatening disease that currently lacks an effective treatment, there will be a higher tolerance for side effects, patient inconvenience, and associated costs. Suppose our new drug is expected to arrest or reverse the pro-gression of Huntington's disease. Patients would likely be willing to accept an increased risk of serious side effects such as cardiac arrhythmias. They would prob-ably also be willing to use the drug even if it required subcutaneous, intravenous, or even intrathecal administration in the doctor's office. And certainly, a drug that pre-vented the death and disability of Huntington's disease could command premium pricing.

Now let us suppose we are developing a novel therapeutic for a less serious condition—a new drug that prevents cataract formation. Since cataract surgery is effective, safe, and quite inexpensive, a new drug must have minimal side effects, convenient oral or topical dosing, and low cost if we expect patients, providers, and payers to support its use. We should also recognize that ophthalmologists might be less likely to champion our drug since it will severely undercut the number of sur-geries they perform.

Box 1.3: What Surprised an Academic?
When we proposed developing a new treatment for prevention of radiation dermatitis (the skin burn that occurs as a result of radiation therapy for malignant tumors), we were surprised when a potential investor insisted that there was no unmet clinical need in this indication. His dermatology expert reported that he never saw patients with this problem. But since dermatologists do not have any effective treatments, radiation oncologists no longer make referrals and instead prescribe emollients to try to alleviate this very debilitating condition. In fact, there is significant unmet need; the burns of radiation dermatitis cause substantial suffering and frequently require withholding further radiation. Our lesson: Cast a wide net—make sure you are speaking with the right experts, and with patients too. —*DM-R*

1.3.4   Feasibility of Development

Our next step is to determine whether it is feasible to develop our new drug. Is there a straightforward clinical development path? What is our target patient population? Are there biomarkers that we can use to predict which patients will benefit? What are the primary and secondary endpoint(s) for clinical trials? How long must we follow the subjects to show efficacy for this endpoint? Are there predictive surrogate endpoints that we can follow? How large is our anticipated effect size? How many subjects must be enrolled? Can we afford to conduct this trial? To answer these questions, we should start with a comprehensive review of the medical literature regarding clinical trials in our indication. We should then speak with physician experts in our chosen disease as well as clinical trial design experts and biostatisticians.

Lastly, we should try to understand the competition in our therapeutic area. What new therapies are in the development pipeline for our chosen indication? What are their mechanisms of action? Do they target the same patient population? If a pharmaceutical company has a two-year head start using our same approach, should we move to another clinical indication or research project? We can explore the competition by doing the following:

1. Search the clinicaltrials.gov website for ongoing clinical trials in our clinical indication
2. Search pharmaceutical industry trade journals for novel drugs in our therapeutic area—these periodicals may be readily available through your university's business school library
3. Search the internet for similar activity
4. Speak with healthcare investors, clinician opinion leaders, and other members of the biotechnology/pharmaceutical community to obtain nonconfidential information about potential competitors

On occasion, we may find it is feasible to develop our drug for several clinical indications. In this case, we should not necessarily pursue the indication with the largest market size. Rather, we should determine which clinical development path has the surest and fastest route to regulatory approval. Once our drug is on the market, we can expand to other indications as part of the "life-cycle management" of the drug.

The Bottom Line

Abraham Lincoln, arguably the greatest leader in the history of the United States, once said, "If I had eight hours to chop down a tree, I'd spend six sharpening my axe." Before spending valuable time and resources executing a new drug development project, we must be certain that:

1. We are advancing an optimized product that addresses the needs of patients.
2. We have a clear path forward.
3. Our approach will still be valued by patients, physicians, and payers when it is finally ready for clinical adoption.

1.4   Target Product Profile

Robert Lum

A Target Product Profile (TPP) is a general term used for a document that summarizes the critical attributes, criteria, and goals that the program or compound needs to meet. TPPs change over the various stages of the drug development process. During the early or research stage, a Candidate Product Profile focuses on what criteria is needed to move the compound forward into animal or IND studies; for a late-stage development program, a Therapeutic Product Profile could have criteria of clinical success and commercial metrics.

At the onset of a project, the criteria can be general, and the TPP is used to guide the overall direction of the project and set "go/no-go" decision points to continue project development. Drug development is a multidisciplinary process. The TPP forces the team to think about attributes outside their area of expertise and how different disciplines influence and change what could be crucial decisions. The more specific the attributes and criteria are defined, the more useful a TPP becomes. A TPP should be used to drive decision-making as to what the critical aspects are for the program and to keep the team focused on the program's ultimate goals.

General characteristics of therapeutics include the clinical indication, route and frequency of administration, medical need, competition, current therapy, cost of intended therapy, stability, clinical development path, regulatory path, and intellectual property (IP) position. This can be a daunting list of categories to consider, but it is important to remember that the team will refine broad characterizations into narrow specification windows as development progresses. SPARK uses a general template to get the process started, shown in Table 1.1.

Table 1.1 SPARK TPP template

Category	Characteristics/attributes	Notes or examples
Product description	Type of agent	Small molecule, peptide, mAb
	Proposed target	Inhibitor of CDK2/CDK4
Indication and usage	Clinical indication(s)	If more than one, specify intended lead indication
	Intended patient population	Females age 18+ diagnosed with HER 2 negative breast cancer
	Current available treatment options	Include surgical, current standard of care, lifestyle, and homeopathic options
Development candidate	Compound characteristics	Molecular weight of <500 No asymmetric centers
	Target specificity	Minimum 10x specificity, CDK2 over CDK4
	Formulation	Tablet, capsules, solution for injection, excipients
	Estimated shelf-life	Greater than 24 months at controlled room temperature, protected from light
	Efficacy	In vitro, cell-based, and in vivo, IC_{50} or EC_{50} for minimum efficacious concentration
Preclinical	Animal model(s) of disease	Accepted "gold standard" or in-house developed model
	Animal toxicity	Single dose, and multiple dose toxicity Two species toxicology studies
	PK/ADME	Minimum half-life, plasma concentrations, excretion Bioavailability for oral administration
	Preclinical safety assessments	Protein binding, hERG, CYP 450, PGP transporters
Clinical Considerations	Trial design	Number of patients, inclusion/exclusion criteria, endpoints
	Dosage and administration	Dosing amount, frequency, etc.
	Route of administration	Oral, IV
Safety/Toxicity in Humans	Known on-target or off-target predicted safety concerns	Known literature or similar compounds in development
	Therapeutic window	Literature precedence or clinical use
Clinical Pharmacology	ADME	Blood brain barrier penetration, primary excreted by kidneys
	Half-life in plasma or serum	Peak and trough concentrations in humans
	PD	Extent of target inhibition or activation
Regulatory Considerations	Presumed clinical path forward	505(b)2, precedents set by previous trials in indication/patient population
	Accelerated development path	Orphan drug status, fast track, breakthrough designation
Intellectual Property	FTO evaluation	Competing patents, opportunities to write new patents
	Composition	Current status of composition of matter, other formulations, process patents

(continued)

Table 1.1 (continued)

Category	Characteristics/attributes	Notes or examples
Financial Considerations	Cost of goods	Buying from pharmacy, cost of goods for drug substance, and drug product
	Competition	Other companies, targets, therapies
	Cost to develop	Size and duration of clinical trials, complexity for development
	Projected pricing and estimated return on investment	Affordability compared to current options. Market penetration, market size, US only, or rest of world

Abbreviations: *ADME* Absorption, Distribution, Metabolism, and Excretion, *CDK* Cyclin-Dependent Kinase, *CYP* 450 Cytochrome P450, *FTO* Freedom to Operate, EC_{50} Half-Maximal Effective Concentration, IC_{50} Half-Maximal Inhibitory Concentration, *HER* Human Epidermal Growth Factor Receptor, *hERG* Human Ether-à-Go-Go-Related Gene, *IV* Intravenous, *PD* Pharmacodynamics, *PK* Pharmacokinetics, *PGP* P-Glycoprotein, *mAb* Monoclonal Antibody, *US* United States

Example 1.1 General TPP for uncomplicated *Falciparum* malaria

General TPP for uncomplicated *Falciparum* malaria		
Development candidate	Formulation	Potential for combination with another agent Pediatric formulation should be available
	Estimated shelf-life	Stable under tropical conditions
	Efficacy	Effective against drug resistant parasites (e.g., those that have developed resistance to chloroquine or sulfadoxine–pyrimethamine treatment) Fast acting and curative within 3 days
Clinical considerations	Dosage and administration	Ideally once, but not more than 3 times per day
	Route of administration	Oral
Intellectual property	FTO evaluation	Requires freedom to operate; composition of matter patent would be ideal
Financial considerations	Cost of goods	Low cost of goods (~US $1 per full course of treatment)

Adapted from Frearson et al. (2007)
Abbreviations: *FTO* Freedom to Operate, *US* United States

Subsequent examples given below are not complete TPPs, but present relevant parts of a profile. Since each project is different, each TPP will have specific criteria that are tailored to each individual development program.

Considering these attributes ahead of time allows the project team to map the path to meet the goals, determine additional expertise that may be needed, and prioritize what needs to get done in the context of the overall program. Example 1.1 provides a brief TPP that defines the general goals of a program.

When developing a new chemical entity, the team uses the TPP to guide their efforts to optimize the characteristics of the lead molecule. The TPP document might include, for example, minimum acceptable criteria for the biochemical assays, cell-based assays, functional assays, selectivity assays, solubility, size (molecular weight), chirality, toxicity profile, formulation, genotoxicity studies, safety pharmacology assays, maximum tolerated dose, efficacy in certain animal models, pharmacokinetic parameters, and IP position. As the program matures, additional criteria may be added, such as pharmacokinetic/pharmacodynamic (PK/PD) relationships, metabolic profiles, frequency of dosing, number of animal models that need to be tested, and additional toxicity studies. The team must define the desired parameters for each attribute. Once all criteria are met, a final set of compounds can be compared and the lead compound selected as a clinical development candidate. Example 1.2 provides a research-oriented TPP, with specific criteria for preclinical testing.

During clinical development, the TPP should be modified to help define more clinically relevant attributes. This includes the primary indication, patient subtypes, dosing regimens, clinical pharmacokinetics, number of patients needed, clinical endpoints, cost of goods, and marketing or commercial strategy. The TPP can also define regulatory strategy, research into companion diagnostics, and alternate

Example 1.2 Hit-to-lead TPP for Protozoa and Helminth Disease

Hit-to-lead TPP for protozoa and helminth disease		
Development candidate	Target specificity	Established selectivity for a molecular target or differential sensitivity between parasite and host enzymes should be >ten-fold
	Efficacy	In vitro activity in antiprotozoan screens: *Plasmodium falciparum*: $IC_{50} < 0.2$ µg/mL *Trypansoma cruzi*: $IC_{50} < 1.0$ µg/mL Antihelminthic screens: *Schistosoma mansoni*: 100% adult worm motility reduction, $IC_{50} < 2$ µg/mL *Onochocerca lienalis, O. ochengi*, or *O. volvulus*: 100% inhibition of micro-filarial motility at 12.5 µM or 10 µg/mL In vivo activity usually in mouse or hamster models: significant reduction in parasitemia and/or increase in life span at a dose of 4 x 50 mg/kg, either through ip or po route, with no overt signs of toxicity
Preclinical	Animal toxicity	Pretoxicity screen in noninfected mice using up to 100 mg/kg ip or po
	Preclinical safety assessments	hERG binding >10 µM Low CYP 450 inhibition profile
	PK/ADME	Metabolic stability determined in microsomes in at least two species, including humans
Intellectual property	FTO evaluation	Should be novel and be able to file for composition of matter patent

Adapted from Nwaka et al. (2009)
Abbreviations: *CYP 450* Cytochrome P450, *FTO* Freedom to Operate, *IC$_{50}$* Half-Maximal Inhibitory Concentration, *hERG* Human Ether-à-Go-Go-Related Gene, *ip* Intraperitoneal, *po* Oral Administration

Example 1.3 Clinical development TPP for a clinical stage glioblastoma cancer drug

Clinical development TPP for a clinical stage glioblastoma cancer drug		
Indication and usage	Intended patient population	Seek approval alone or in combination with bevacizumab for treatment of glioblastoma multiforme, which has progressed after treatment with radiation plus temozolomide
Clinical considerations	Trial design	Median PFS >6.3 months compared with 4.2 months for bevacizumab alone Median overall survival >9 months for combination
	Dosage and administration	120 mg/m^2 IV once every 3 weeks until disease progression or 6 cycles
Safety/toxicity in humans	Known on-target or off-target predicted safety concerns	Grade 3 or 4 neutropenia assumed in majority of patients; manageable with growth factor support Neuropathy Grade 3 or 4 in <10% of patients Other toxicities manageable, predictable, and reversible
Intellectual property	FTO evaluation	Seek patent protection for novel combination therapy with bevacizumab
Financial considerations	Cost of goods	Sustainable supply chain with cost of goods: <$50 per vial

Adapted from unpublished program
Abbreviations: *FTO* Freedom to Operate, *IV* Intravenous, *PFS* Progression Free Survival

therapeutic indications or formulations. Example 1.3 provides a TPP for a compound in clinical development that may be used to guide the team during the clinical development phase.

Box 1.4: What Surprised an Academic?

When I co-founded KAI and joined the company for its first year, I was frustrated at first when the VP of Drug Development organized a set of meetings to create an explicit TPP. I did not understand the need for such meetings; after all, we knew where we were going. Why waste time stating the obvious? I quickly learned how critical this process is. We need to plan with the end in mind—a mantra that we keep repeating at SPARK. Drug development is highly interdisciplinary, and defining important characteristics by developing a TPP has proven to be essential; it mapped our path, identified whom we needed to engage, and established optimal attributes for our product. —DM-R

The Bottom Line

The TPP is a living document that allows you to develop your projects, **starting with the end in mind**. The TPP should define the desired attributes of the novel therapeutic under development and should be edited and refined as the product moves further through the development pipeline. An effective TPP includes: clinical indication and medical need, route and frequency of administration, current and future competition, cost of intended therapy, IP position, and all other advantages over current treatments. Other possible attributes include clinical development path, regulatory path, and metabolic and safety profiles.

1.5 Project Management and Project Planning

Rebecca Begley and Daria Mochly-Rosen

Most members of academic research teams are trainees with expertise in similar disciplines, led by a principal investigator who sets the research agenda while ensuring the continual education of the junior scientists in the lab. As a result, research is not closely tracked against a formal timeline and the research plan can rapidly change direction to pursue new and interesting observations; little attention is given to actively manage, coordinate, and adhere to a timeline for the research enterprise.

In industry, coordination of efforts across a wide range of disciplines, tracking progress, and adhering to a plan are essential. Therefore, project management is a highly valued function that substantially increases the likelihood of a successful outcome and saves both time and money. Project teams bring together individuals with varying levels of seniority and widely divergent areas of expertise, such as pharmacology, toxicology, regulatory science, drug manufacturing, and clinical trial design (referred to as cross-functional teams). Team members are committed to advancing their project in a timely and collaborative manner. They are also encouraged to kill a project as soon as possible if the research indicates that the project is unlikely to succeed or will incur unacceptable costs or delays.

1.5.1 Project Leadership

Project management requires strong leadership, a committed team with the necessary complement of skills, and a well-thought-out development plan. The project team works together to identify the project's strategy (vision), goals (tactics), and a

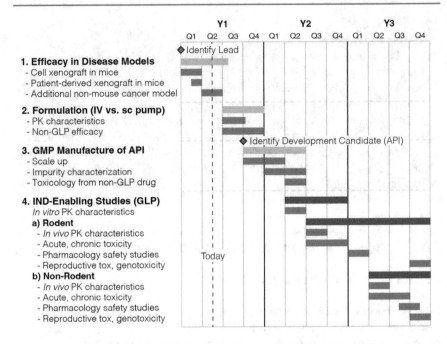

Fig. 1.4 Theoretical Gantt Chart for a preclinical-stage oncology program. This project Gantt Chart outlines possible tasks and predicted timelines for a hypothetical development program. Abbreviations: *API* Active Pharmaceutical Ingredient, *GLP* Good Laboratory Practice, *IV* Intravenous, *IND* Investigational New Drug Application, *PK* Pharmacokinetic, *Q* Quarter, *SC* Subcutaneous, *tox* Toxicology, *Y* Year

detailed execution plan. The project leader then helps keep the team on task according to a predetermined budget and timeline and works with the team to find acceptable solutions to inevitable challenges that arise.

Importantly, many of the tasks carried out by team members are highly interdependent. For example, manufacturing the drug supply for a clinical study cannot begin until the appropriate clinical dosing regimen has been determined by the clinical team member. Clinical dosing for a first-in-human trial is furthermore highly dependent upon toxicology, pharmacokinetics, and efficacy parameters characterized in animal studies that are conducted by preclinical experts on the team.

A Gantt chart is a useful tool that provides a detailed road map for executing and tracking the development project. It includes a comprehensive listing of each task that must be accomplished during the project, along with its anticipated timeline and its dependencies upon other parts of the project (Fig. 1.4). The project manager can use the Gantt chart to track the progress of each task as well as overall progress of the project against the desired timeline. Similarly, team members representing different functional areas can track their tasks and see how slippage in their timing might affect the overall timeline. For example, a delay in delivery of acceptable quality drug product will delay the start of IND-enabling toxicology studies and, in turn, delay filing of the IND. That may seem obvious, but such a delay may result in losing a time slot at the CRO conducting the toxicology studies, which would cost the company in penalty payments to the CRO and further delay development. The costs and consequences of small delays can quickly add up in drug development.

The Gantt chart is not set in stone and should be revised by the project manager as new information becomes available. Although complications invariably arise during drug development, the Gantt chart is an extremely useful instrument to help the development team complete the project on a timeline that fits the company's goals.

Box 1.5: What Surprised an Academic?

A Gantt chart is rarely used in academic research to identify specific goals and track progress towards them. Who can plan basic research with such detail? When asked to participate in this planning, I felt it was a waste of time. I quickly realized that such detailed planning is an effective tool to create priorities, to know when to "kill a project" (e.g., it will be completed too late to impact the company's future, or the technical setback is so substantial that it is too expensive to complete), how to keep the project moving on track, and how to take corrective actions when budgets and/or timelines change. —DM-R

1.5.2 Project Management for SPARK

We include a section on project management because academics engaged in translational research must take on this function to ensure timely and successful completion of their aims. The project leader may be the faculty member or can be a student or a fellow. The team may include expert advisors, other research laboratories at the same institution or elsewhere, as well as commercial research services (e.g., medicinal chemistry or toxicology). The team members in this case are not bound in the same way that they are in a lab or in a typical project team in a company. Further, it is unlikely that the project leader will be able to assemble all function heads for a meeting; therefore, a lot of project planning relies on coordination and individual conversations with each expert and function. When possible, it is advisable to share plan details with all members of the team to confirm assumptions and coordinate progression. The following section and suggested references provide some practical advice on leading cross-functional teams; not all of it may apply to academic work (Kennedy 2008; Linberg 2006).

Box 1.6: What Surprised an Academic?

Much to our surprise, most SPARK project failures have been the result of human rather than scientific factors. For example, a critical postdoctoral fellow may have left the university, causing the project to languish for lack of a champion. Less commonly, the team failed to execute in a timely manner or would not follow enabling advice from experienced advisors. These failures are particularly frustrating, because the underlying science remains promising, but is unlikely to help patients. The importance of managing time and human resources in translational research cannot be overstated. —KVG

1.5.3 Leading a Cross-Functional Team

How to lead when you are not the expert or the most senior person in the team:

- Influence without authority depends on relationships and shared vision. Build the relationships before you need them.
- Stay flexible; adjust to new data or changes in circumstances.
- Know enough about each functional area's activities to converse intelligently. You should understand where key issues may arise and why. Ask questions early; establish mentors/go-to people to gain basic understanding.
- Use the cross-functional team meetings as a forum for holding the entire team accountable to the project and each other.
- Use cross-functional team meetings to identify and address issues that arise from within each function as well as from an interface with another function (as in the example above, on the consequences of delay in production of drug supply).
- When contentious issues arise, conducting individual discussions with key stakeholders can help resolve issues in advance of the cross-functional meeting.
- Written documentation can be useful for team management. Writing down goals, targets, and decisions provides a common point of reference for communication, both internally within the team, as well as to external audiences. In addition, gaining team agreement on a written document can encourage more attention to the wording (written agreements can carry more weight than spoken ones) and can facilitate a greater degree of group buy-in if the group feels involved in the process.
- Tools for communication include the TPP and Gantt chart. These will likely evolve over time.
- Communicate, communicate, communicate. Engage stakeholders early and often. Ensure satellite discussions and decisions are brought back to the team.

1.5.4 Aspects of Project Planning

Step 1: Plan with the end in mind—define the vision of the project.

The project plan should be determined starting with the final product and working backwards. We begin with defining a target product profile (TPP) with the team. The TPP describes our final product and why a physician or patient would use it by highlighting where it addresses unmet need. The "must-have" characteristics outlined in the TPP define the threshold below which the project would not be carried forward and thus should be "killed." Published clinical trial data and product labels are resources for comparative information on related products. (See Sect. 1.4 for more information on TPPs.)

Step 2: Outline a clinical development plan.

The clinical development plan can then be structured by using the TPP and working backwards to the current stage of development. A discussion with clinicians and business development advisors should help determine the development plan. This is important even for an early-stage project, and broad descriptions will suffice. The development plan outlines decision points in the overall project and details activities needed to advance from the current state to the next decision-making point. In addition, key risks and assumptions for the project are summarized.

After identifying the desired clinical indication in the TPP, we should design the phase 3 clinical trial required for regulatory approval and clinical adoption. We can then determine what the preceding phase 2 and phase 1 studies would have to look like to support dose selection and study design for the phase 3 trial. These discussions should include variables such as endpoints, duration of treatment, number of doses, and size of study. As these are likely to change as the program evolves, test the boundaries of the proposed numbers. For example, if the clinician recommends that we treat this patient population for 1 month to observe a significant change in a particular endpoint, we should query how likely it is that we would end up treating for 2 months or if it would be feasible to treat for 2 weeks instead. The rest of the team (toxicology, manufacturing, pharmacology, etc.) should be asked to propose activities that would be needed from their areas to support the clinical program as described. These activities should answer "key questions" that exist for the project.

Step 3: Lay out the project plan with all details to facilitate decision-making.

Once the clinical development plan is placed into a timeline with an accompanying budget, it is time to determine the preclinical development plan. What GLP studies will be required before filing an IND? What additional non-GLP (in vitro and in vivo) efficacy, PK/ADME, and preclinical safety studies should be conducted prior to a pre-IND meeting with FDA? Document the assumptions used to pull together this plan and ensure all envisioned activities needed to support the project are included. (While such planning may seem to be excessive for a program conducted within academia, having a thought-out and detailed plan for your product all the way through phase 3 will help secure licensing or VC funding.)

Step 4: Define the activities needed to reach the next decision-making point and set goals accordingly.

As we review the overall project plan, inclusive of all proposed activities, we can prioritize the activities and determine which will add the most value to the project upon completion. For example, conducting a GLP safety study will add some value, but completing two independent non-GLP efficacy studies in animals may provide greater value to the project. Of course, proposed activities should be weighed against the available budget.

The focus and priority should be on the must-have activities. These essential steps become the project plan. The plan should be revisited upon receipt of new data or a change in the project environment (e.g., approval of a new agent in the disease indication).

The Bottom Line

Even while in academia, a development plan is an essential map for the team, navigating us through the many interdependent processes of drug development and defining critical "go/no-go" decision points to continue or terminate the project. The plan is a living document without which we may wander off task, waste precious resources, and create delays in reaching our goal: to benefit patients. Additionally, a good and detailed plan will help secure VC funding and partnering.

Key Terms and Abbreviations

Key Terms

Biologics License Application (BLA): FDA paperwork to obtain approval for the sales and marketing of a new biologic in the US.

Cross-Functional Team: Project team comprised of individuals with expertise in different areas (e.g., pharmacology, ADME, manufacturing, regulatory science, clinical) required for successful completion of the project.

Development Candidate: Molecule selected for clinical development after meeting criteria established in the TPP for efficacy, pharmacokinetics, and safety; may subsequently be called drug substance or active pharmaceutical ingredient (API).

Drug Product: Active pharmaceutical ingredient and inactive components such as binders, capsule that compose the final drug formulation.

Excipient: Inactive material added to the formulation to control drug dissolution, absorption, stability, etc.

Gantt Chart: Development plan tracking tool listing critical tasks, timelines, and dependencies.

Good Laboratory Practice (GLP): Extensive documentation of each procedural step to ensure high-quality, reproducible studies.

Good Manufacturing Practice (GMP): Exacting procedures and documentation of quality assurance carried out at a certified facility (sometimes referred to as "cGMP" for "current" practice).

Hit: Molecules that display the desired activity in an assay.

IC_{50}: Drug concentration required to inhibit a process by half its full activity.

Institutional Review Board (IRB): The entity designated to review and monitor research involving human participants in the US; often called the Ethics Committee outside the US.

Investigational New Drug Application (IND): Document filed with FDA prior to initiating research on human subjects using any drug that has not been previously approved for the proposed clinical indication, dosing regimen, or patient population.

Lead: Most promising early-stage molecule(s) identified through in vitro and in vivo testing.

New Drug Application (NDA): FDA paperwork to obtain approval for the sales and marketing of a new drug in the US.

Office of Technology Licensing (OTL): The university group responsible for managing intellectual property.

Omics: A branch of science dealing with large-scale, quantifiable, datasets that aim to capture a nonbiased comprehensive view of a biological system (e.g., genomics, metabolomics, proteomics, transcriptomics). "Omics" have become a major source of target identification in drug discovery.

Pre-IND Meeting: Meeting with FDA prior to submission of the IND to provide guidance on data necessary to warrant IND submission; ideally occurs prior to initiating expensive GLP preclinical studies.

Repurposed Drug: An FDA-approved drug with a new indication, formulation, or route of administration.

Target Product Profile (TPP): A document outlining essential characteristics of the final drug product.

US Food and Drug Administration (FDA): A US federal agency responsible for protecting public health by assuring the safety and efficacy of drugs, biological products, medical devices, and other products.

Key Abbreviations

ADME	Absorption, Distribution, Metabolism, and Excretion
API	Active Pharmaceutical Ingredient
AI	Artificial Intelligence
CMS	Centers for Medicare & Medicaid Services
CMC	Chemistry, Manufacturing, and Controls
CRO	Contract Research Organization
HTS	High-Throughput Screening
hERG	Human Ether-à-Go-Go-Related Gene
IPO	Initial Public Offering
IP	Intellectual Property
ip	Intraperitoneal
mRNA	Messenger RNA
miRNA	MicroRNA
mAb	Monoclonal Antibody
po	Oral Administration

PD	Pharmacodynamics
PK	Pharmacokinetics
PFS	Progression Free Survival
PKC	Protein Kinase C
QA	Quality Assurance
SSRI	Selective Serotonin Reuptake Inhibitor
SI	Sensitivity Index
SNRI	Serotonin–Norepinephrine Reuptake Inhibitor
siRNA	Small Interfering RNA
SAR	Structure–Activity Relationship
VC	Venture Capital

References

Etzkowitz H, Mack A, Schaffer T, Scopa J, Guo L, Pospelova T (2020) Innovation by design: SPARK and the overcoming of Stanford University's translational "valley of death" in biomedicine. Manag Decis Econ 41:1113–1125

Frearson JA, Wyatt PG, Gilbert IH, Fairlamb AH (2007) Target assessment for antiparasitic drug discovery. Trends Parasitol 23:589–595

Gehr S, Garner CC (2016) Rescuing the lost in translation. Cell 165:765–770

Kennedy A (2008) Pharmaceutical project management. In: Drugs and the pharmaceutical sciences, vol 182. Informa Healthcare, New York

Kim ES, Omura PMC, Lo AW (2017) Accelerating biomedical innovation: a case study of the SPARK program at Stanford University, School of Medicine. Drug Discov Today 22:1064–1068

Linberg SE (2006) Expediting drugs and biologics development: a strategic approach.

Nwaka S, Ramirez B, Brun R, Maes L, Douglas F, Ridley R (2009) Advancing drug innovation for neglected diseases—criteria for lead progression. PLoS Negl Trop Dis 3:e440

Therapeutics and Diagnostics Discovery

2

Daria Mochly-Rosen, Kevin Grimes, Rami N. Hannoush,
Bruce Koch, Gretchen Ehrenkaufer, Daniel A. Erlanson,
Julie Saiki, Jennifer L. Wilson, Shelley Force Aldred,
Adriana A. Garcia, Jin Billy Li, Rosa Bacchetta,
Maria Grazia Roncarolo, Alma-Martina Cepika,
Harry Greenberg, Steven N. Goodman,
and Michael A. Kohn

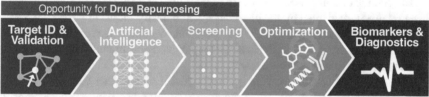

Daria Mochly-Rosen, Ed, Kevin Grimes, Ed.

The original version of the chapter has been revised. A correction to this chapter can be found at
https://doi.org/10.1007/978-3-031-34724-5_7

D. Mochly-Rosen (✉) · K. Grimes
Chemical and Systems Biology, Stanford University School of Medicine, Stanford, CA, US
e-mail: sparkmed@stanford.edu; kgrimes@stanford.edu

R. N. Hannoush
EpiBiologics, San Mateo, CA, US

B. Koch
Sarafan ChEM-H and Stanford Innovative Medicines Accelerator, Stanford University,
Stanford, CA, US

G. Ehrenkaufer · A. A. Garcia
SPARK at Stanford, Stanford University School of Medicine, Stanford, CA, US

D. A. Erlanson
Frontier Medicines, South San Francisco, CA, US

J. L. Saiki
SPARK at Stanford Advisor, Stanford, CA, US

J. L. Wilson
Department of Bioengineering, University of California Los Angeles, Los Angeles, CA, US

In any therapeutics discovery and development effort, multiple critical steps must be accomplished to arrive at a compound that is safe and efficacious, while also exhibiting the complex array of desired drug-like behaviors that warrant advancement to the clinic. These tasks include target identification and validation, screening for active compounds, chemical modification of candidate compounds to achieve optimized pharmacology, formulating the final drug product, and establishing safety in preclinical models. "Repurposing" drugs that have previously been approved (or shown to be safe in humans) for new clinical indications can provide a faster, less risky, and more cost-effective route to bring a new therapy to patients. Such shortcuts in development can be particularly valuable to resource-constrained academics.

In addition to small molecule therapeutics, new drug discovery tools and novel drug modalities have evolved in recent years; many of them made their first steps in academia. These include integration of artificial intelligence (AI) in identifying and evaluating drugs and drug targets and therapeutic modalities such as antibody, RNA interference, DNA and RNA editing, and cell therapy.

Despite the urgent needs, fields such as antibiotic and vaccine development have been relatively neglected by the pharmaceutical industry in recent years (see Sects. 6.1 and 2.10, respectively). The COVID pandemic has illustrated the importance of continued research into these fields and demonstrated how quickly advances can be made when prioritized appropriately.

Finally, diagnostic discovery has gained renewed interest, not only as important tools for clinical practice, but also in the process of drug discovery and development. Diagnostic biomarkers can help reduce the cost of drug discovery by providing evidence of a drug's pharmacodynamic effect, serving as clinical trial surrogate endpoints that can predict long-term clinical benefit, and identifying those patients most likely to benefit from a given therapeutic approach (a field now termed personalized medicine).

This chapter briefly describes the discovery and use of various modalities of therapeutics and diagnostics and how they can be used to address common and rare diseases.

S. F. Aldred
Rondo Therapeutics, Hayward, CA, US

J. B. Li
Department of Genetics, Stanford University School of Medicine, Stanford, CA, US

R. Bacchetta · M. G. Roncarolo · A.-M. Cepika
Department of Pediatrics, Division of Hematology, Oncology, Stem Cell Transplantation and Regenerative Medicine, Stanford University School of Medicine, Stanford, CA, US

H. Greenberg
Department of Medicine/Gastroenterology and Hepatology, Stanford University School of Medicine, Stanford, CA, US

S. N. Goodman
Department of Epidemiology and Population Health, Stanford University School of Medicine, Stanford University, Stanford, CA, US

M. A. Kohn
Department of Epidemiology & Biostatistics, University of California San Francisco (UCSF) School of Medicine, San Francisco, CA, US

2.1 Finding and Validating Drug Targets

Rami N. Hannoush and Daria Mochly-Rosen

One of the most common outcomes of academic research is identifying molecular events that contribute to a particular human pathology. Our basic research to understand how the cell/organism works uses biochemical, imaging, and genetic tools to confirm that a particular molecular event is required and sufficient for the effect we describe. These Robert Koch's postulates of causality, published in 1890 (Koch 1890), drive much of biological research; we often use gain- and loss-of-function approaches to confirm that a particular protein of interest is indeed responsible for the cellular process that we study.

Gain- and loss-of-function approaches also apply to finding and validating drug targets, essential processes in drug discovery. For example, type I diabetes, characterized by rapid weight loss and high blood (and urine) glucose levels, is caused by lack of insulin. Experiments in 1921 showed that occluding pancreatic ducts in dogs caused diabetic symptoms, and these symptoms were alleviated by injecting pancreatic extract, or purified insulin. In other words, removal of the pancreas (and the insulin it produced) was required and sufficient to produce diabetes. Within 2 years, this was confirmed in humans: In 1923, pancreatic extract injected into a 13-year-old boy on the brink of death from diabetes had spectacular success; his blood and urinary sugars returned to normal and other diabetic symptoms were alleviated. Although the above approach linked lack of insulin to type I diabetes, it did not elucidate the reasons for the low insulin levels. Other important targets that might prevent or reduce the burden of this pathology were not identified for many years.

To elucidate important contributors to diseases, we now recognize that any human disease can be represented as a network, correlating symptoms with underlying cellular processes. Current correlation maps of human diseases focus mainly on genetic variations and levels of gene expression. More recently, these maps include other "omics"—characterization tools such as epigenomics, proteomics, metabolomics, microbiomics. These data help link genetic mutations and polymorphisms to downstream effects on RNA and proteins that cause the human phenotypes that comprise a disease. Interpreting these maps allows us to identify biomarkers associated with the disease of interest and potential therapeutic targets.

A similar approach can be used to confirm the validity of disease models and importantly, to identify which elements correlate with disease progression and predict drug benefit in the later phase of drug testing. Indeed, target identification constitutes a more intuitive effort for academic researchers: It relies on early stage research in the laboratory that links a target with a disease. This connection between target and disease can be further strengthened by scientific reports from other laboratories, human genetic studies such as genome-wide association study (GWAS) analysis, expression levels of the target (functional genomics), and proteomics. Studies using human samples can be especially compelling.

After identifying a target, the next essential step is validation of the target. This is a highly critical step in drug development and includes several separate activities. The first step is to confirm the causal relationship between the protein and the phenotype, applying Koch's postulates. Here both genetics and pharmacology tools can be used. For example, we presume that a particular protein causes the disease/phenotype if its overexpression leads to disease and its knockout prevents the phenotype. However, a protein may be required to prevent a pathology, without its loss being sufficient to cause it. In that case, the disease phenotype may require another trigger; over-expression or knockout of another gene (or pharmacological agents that activate or inhibit that protein, when available) may induce a phenotype only in the presence of another gene or an environmental factor. For example, inhibiting an enzyme that reduces oxidative stress, and therefore reduces a certain pathology, will have no phenotype if no oxidative stress is present.

Second, validation studies need to be conducted in several relevant model systems of the disease. These models should include cultured cell lines in which modulating the target induces a cellular phenotype that correlates with the disease phenotype. An effective model system uses cultured patient-derived cells or, through production of induced pluripotent cells, their differentiation to the cell type relevant to the disease. In vivo models in rodents or other species are particularly important, even though in many cases such models are imprecise predictors of the human disease.

A third means to validate the target is the identification of human biomarkers that correlate with disease progression collected from a large patient population. These biomarkers may identify patient subpopulations with the same disease subtype. Importantly, we need to show that the patients' biomarkers are recapitulated in at least some of the animal models. Means for identifying, selecting, and validating drug targets are summarized in Fig. 2.1.

Once the molecular target has been validated, the selection of the "hit" compounds (described in the following sections of this chapter) that modulate the target will use similar validation approaches, although often fewer variables are included. Therefore, the success of your drug development effort depends on identifying the correct molecular target that is implicated in a human pathology and, importantly, the correct disease models based on a thorough target validation. Solid and independent target validations will increase the likelihood of successfully identifying pharma partners and/or investors. More importantly, target validation will allow you to map effective go/no-go criteria for your drug development efforts: The models that were used to validate the drug target are the models that you will likely use to test your intervention, such as hits from screening campaigns. As discussed later in this book (Sect. 3.2), the disease models that you plan to use should be robust, easy to reproduce by independent researchers, and hopefully, models that have already been successfully used by others.

The Bottom Line

Multiple approaches should be integrated to identify and validate the target and eventually, confirm the intervention. Common approaches include knocking down the expression levels, overexpression of the molecular target, identifying genetic variants in humans that affect disease risk or progression, the use of genetic gain- or loss-of-function studies in tissue culture and in animals, the use of patient-derived cells in culture, and the use of patient-derived biomarkers associated with the disease of interest.

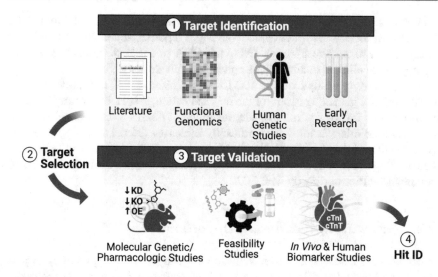

Fig. 2.1 Target identification and validation. Potential targets may be identified via a variety of approaches, including literature reviews, functional genomics studies, human genetic studies, and early research. After a target has been identified and selected, the target must be validated. This includes proof-of-concept studies using genetic and pharmacologic approaches, feasibility studies to ensure manufacturing of drug is possible, and performing biomarker studies. Abbreviations: *ID* Identification, *KD* Knockdown, *KO* Knockout, *OE* Overexpression. (Created with BioRender.com)

2.2   Repurposing Drugs

Kevin Grimes

Drug repurposing (also called drug repositioning) refers to the practice of developing an existing drug for a new clinical indication. Ideally, a drug selected for repurposing has been tested extensively in humans and has a known safety profile. The drug may have received regulatory approval for its original indication or may have stalled in development, perhaps due to lack of efficacy or an unacceptable toxicity profile for a nonserious clinical indication.

Repurposing can be a faster, less risky, and more cost-effective route to benefit patients and is therefore particularly attractive for academics and other not-for-profit drug developers. Pharmaceutical companies, biotechnology companies, and healthcare investors are often less enthusiastic about supporting the development of a repurposed drug because the active compound is typically not patentable. Nonetheless, proprietary claims regarding formulation, dosing, or clinical indication may allow a period of exclusive marketing and lead to a profitable program. For instance, repurposing the teratogenic sedative thalidomide to treat multiple myeloma is an example of the profitable exploitation of a drug whose patent had long ago expired.

While physicians often prescribe drugs for "off label" uses when caring for individual patients, a drug repurposing development program for a novel indication requires clinical human experimentation and, therefore, approval from your Institutional Review Board (IRB). Advancing a repurposed compound to clinical study may also require filing an IND with FDA or relevant national regulatory agency (if the clinical studies will be conducted outside the US).

Drug studies typically require a new IND if the research will be reported to FDA in support of a marketing claim for the new indication, that is, a new drug label, or if the research involves a "route of administration or dosage level or use in a patient population or other factor that significantly increases the risks (or decreases the acceptability of the risks) associated with the use of the drug product" (Code of Federal Regulations, 2023). When in doubt, check with your institution's legal or compliance office or directly with FDA.

2.2.1 Identifying Repurposing Opportunities

When we have discovered a novel, validated drug target, screening a library of previously approved drugs for activity against our target may lead to a drug repurposing opportunity. Researchers at the US National Institutes of Health (NIH) have assembled a comprehensive list of drugs that have been previously approved by FDA ($n = 2356$) and by regulatory agencies worldwide ($n = 3936$, inclusive of FDA). In addition, NIH has compiled a library of 2750 of these previously approved drugs and of 4881 drugs that have undergone human testing, but have not been granted regulatory approval (Huang et al. 2011). Researchers may apply to have NIH test their targets against this library. Alternatively, many high-throughput screening (HTS) centers now also include a collection of previously approved drugs as a part of their chemical library. HTS is discussed in greater detail in the next section.

A second path to repurposing is to apply a known modulator of a specific biologic target to a new disease. For example, eflornithine is an inhibitor of ornithine decarboxylase (ODC), a key enzyme in mammalian cells for converting ornithine to polyamines. The polyamines, in turn, are important in cell proliferation, differentiation, and growth. Eflornithine stalled in development when it failed to show adequate efficacy as an antitumor agent, but has subsequently been successfully redirected as a treatment for African sleeping sickness, since ODC is also present in the causative parasite.

A third avenue for identifying repurposing opportunities is through astute clinical observation and exploitation of known or unanticipated side effects. For example, erythromycin is well known for causing gastrointestinal distress and diarrhea. This observation led to its clinical use as a promotility agent in selected patients with a functional, nonobstructive ileus. Similarly, sildenafil originally entered clinical development as an anti-angina/antihypertensive agent. A serendipitous clinical observation led to its development as a treatment for erectile dysfunction—an extremely lucrative market opportunity.

A fourth repurposing approach is to use artificial intelligence to identify repurposed compounds. For example, a computational algorithm may be utilized to

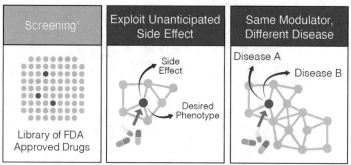

Fig. 2.2 Strategies for identifying repurposed drugs. Various strategies can be used to identify drugs for repurposing. This includes screening against a library of FDA-approved drugs, exploiting unanticipated side effects of a preexisting drug, or repurposing a drug that modulates a target involved in multiple diseases. Abbreviations: *AI* Artificial Intelligence, *FDA* US Food and Drug Administration

search for structural homology between a tool compound and previously approved drugs. The "hits" can then be tested in the same preclinical assays for the desired activity. Machine learning can also be applied to identify protein targets downstream of the validated target. If drugs exist for these adjacent targets, they may also produce the desired effect. Computational analysis may also be utilized to identify drugs that modulate gene expression in the opposite direction of your disease of interest. Lastly, searching large clinical databases for improved outcomes in patients who are exposed to drugs commonly prescribed for other conditions may lead to new drug-target combinations to explore. Strategies for identifying repurposed drugs are summarized in Fig. 2.2. The use of AI is described in greater detail in Sect. 2.6.

The following sections will discuss the repurposing of drugs based upon the drug's regulatory status, patent status, and intended indication, dose, and route of indication. In general, regulatory agencies will focus first and foremost on the safety of the proposed dosage and formulation in the new patient population. Of course, we must also show efficacy to gain regulatory marketing approval. Strategies for developing repurposed drugs are summarized in Fig. 2.3 (this is a general guideline, and it is recommended that you consult with someone experienced in drug repurposing).

2.2.2 Previously Approved Drugs Using the Same (or Lower) Dose and Route of Administration

This category presents the fastest route to the clinic. If the drug is generically available and the intended patient population is not at increased safety risk, there are relatively few barriers to conducting a clinical study and publishing the results. Of course, IRB approval is required prior to initiating the study. Once the results are

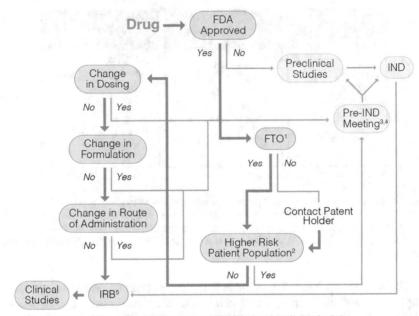

Fig. 2.3 Strategies for developing repurposed drugs. Once a candidate has been selected for repurposing, several key questions (illustrated as a decision tree) may guide your repurposing strategy. Abbreviations: *FTO* Freedom to Operate, *IRB* Institutional Review Board, *IND* Investigational New Drug Application, *FDA* US Food and Drug Administration

published, physicians will be free to prescribe the drug off-label without a formal regulatory approval for the new indication. If there is reason to suspect increased risk or that the known drug risks are less acceptable for the intended indication and study population, we must file an IND. In general, it is good practice to check with FDA before assuming that an IND will not be required.

Importantly, there are significant barriers to procuring adequate funding for clinical development of a repurposed generic drug. Currently, most of these projects stall for lack of funds, no matter how promising. Without patent protection, there is little opportunity to generate a return on investment, which prevents the biopharmaceutical industry from participating. Unfortunately, nonprofit drug development is nearly nonexistent. Federal grant funding for clinical trials is quite limited and often highly proscribed. Clearly, we need novel funding approaches to support the clinical development of generic drugs for novel indications. Successful development might not only improve patient outcomes, but substantially lower the cost of treatment.

If the drug is proprietary, we should consider approaching the company that markets the drug to solicit support for our study. Depending upon the size of the current market and the number of years remaining on the patents, the company may see our repurposing proposal as either an opportunity or a threat. Our proposed new market may represent an attractive pipeline extension. On the other hand, unanticipated negative adverse effects in the clinical study may threaten the existing franchise. If an IND is required, we must have the company's approval for FDA to access its proprietary Drug Master File at the agency; thus, company consent is required. If an IND is not required, we may proceed with our study, even without the company's consent, assuming that we have obtained IRB approval and have adequate financial resources.

Working with the company can provide many advantages beyond financial support or free study drug. The company scientists will have an extensive working knowledge of the drug's metabolism, formulation, side effects, and potential drug–drug interactions. This information can be invaluable in the design and execution of the new clinical study.

Box 2.1: What Surprised an Academic?

Dr. Al Lane, a Stanford pediatric dermatologist, cared for a child with a rare, deforming, and sometimes lethal pediatric disease, called lymphatic malformation. When his patient was treated with sildenafil for pulmonary arterial hypertension, he noted that the large mass within the child's chest dramatically decreased in size over a period of several months. He recognized the therapeutic potential for this previously untreatable condition and requested IRB approval to administer the drug to a second patient, who also showed a dramatic response. Dr. Lane applied to SPARK in the hopes of conducting a small corroboratory clinical study, which was enabled when Pfizer generously donated the study drug. After promising results in seven additional children, Dr. Lane received a substantial grant from FDA to support a definitive clinical trial. We were surprised that FDA was the source of the grant, but the agency does provide funding to support clinical trials for rare diseases. The publication of these initial cases has impacted clinical practice—an outstanding outcome considering the lack of financial incentive. Dr. Lane's serendipitous observation and ability to find funding from a nonobvious source are a tribute to his clinical prowess and perseverance. In the case of non-profit drug development, government agencies and other non-profit organizations can provide critical financially assistance and even help in recruiting patients. Cast a wide net. —DM-R

2.2.3   New Route of Administration, Dosing, or Formulation

Regulatory agencies require that a drug be both safe and efficacious. When a drug is administered via a different route (*e.g.*, via inhalation instead of intravenously), at higher dosages, or in a new formulation, the safety profile will be altered, and human efficacy will be unproven. Therefore, an IND is required.

Although prior human experience with the drug can be predictive and help guide preclinical studies, supplemental Good Laboratory Practice (GLP) safety studies will typically be required to determine that the route, dose, or formulation is safe to test in humans. At a minimum, preclinical studies should be conducted to assess safety and characterize pharmacokinetics for the new formulation and/or route of administration. Non-GLP preclinical efficacy studies can be useful in demonstrating biological effect and predicting the clinical dosing requirements. An open discussion with the regulatory agency early during development can be invaluable in determining which preclinical studies will be required prior to entering clinical study.

2.2.4 Nonapproved Drug with Human Trial Data

A number of drugs fail to advance beyond their initial phase 2 or 3 clinical study due to lack of efficacy for their original intended clinical indication. These drugs are typically "shelved" by the sponsoring company but can be very valuable if a new target or clinical indication can be identified. The timeline for developing a "shelved" drug for a new indication can be appreciably shortened and less costly because the company sponsor already has a complete preclinical package, human safety data, and a Drug Master File with FDA (or similar regulatory agency). Often, clinical grade drug product is also available if it still meets its quality specifications. Typically, we must work with the original company sponsor because the drug is under patent protection and/or all the previous data filed with the regulatory agency is proprietary and owned by the company. The National Center for Advancing Translational Sciences (NCATS) at NIH offers a New Therapeutic Uses program for academic researchers that provides access to libraries of drugs that have stalled in development (National Center for Advancing Translational Sciences 2022).

> **The Bottom Line**
> Drug repositioning can be a faster, less risky, and less expensive route to develop a new therapy for a clinical indication. Repurposing is particularly attractive for academics and other not-for-profit drug developers who are seeking cures for patients but have limited financial resources. Some repurposing programs can be quite successful commercially if they have intellectual property claims that block competitors or privileged regulatory status (*e.g.*, orphan disease designation).

2.3   Developing Assays for High-Throughput Screening

Bruce Koch

The aim of high-throughput screening (HTS) of chemical libraries is to identify small molecules (chemical leads) that "hit" or affect a protein target or cellular phenotype. The screen typically identifies good starting chemical entities that will be improved upon using medicinal chemistry (optimization).

Alternative approaches exist for identifying small libraries of chemical leads, such as searching the published literature (including patents) or screening substrate or transition state analogs; however, these approaches only apply to a subset of drug targets. In silico and fragment-based screening are also options for exploring large areas of chemical space, but these methods require prior target structure elucidation and, in the case of fragment-based screening, high assay sensitivity. (See Sect. 2.6 for further guidance on informatics-based searches.) This section focuses on the development of activity-based assays for identifying and characterizing compounds from large (>100,000 compounds), drug-like molecule libraries arrayed in microtiter plates with one compound per well using HTS, which is a source of leads for many drug discovery programs.

HTS relies on robust, miniaturized, "mix and measure" (does not require washing away components) assays. A robust assay is one with a high Z'-factor (defined in Box 2.2)—a measure of the assay signal relative to noise (Zhang et al. 1999), good reproducibility between runs, and resistance to interference. With the large compound libraries typically screened via HTS, cost and logistics often dictate that only a single well per compound be run. Thus, even with a high Z'-factor, there are considerable opportunities for false-positive (nonreproducible) and false-negative (missed actives) results due to random variation. These factors should be considered when designing, optimizing, and characterizing the primary screening assay. Often, multiple iterations of assay design and testing are required to adapt a low-throughput (<50 samples) assay for optimal performance in HTS.

Box 2.2: Z'-Factor Defined

The Z'-factor reports the window between an assay's signal (positive control) and background (negative control) relative to assay noise. Good HTS assays have a Z'-factor between 0.5 and 1. A higher Z'-factor indicates a larger difference between the assay signal and background noise.

$$Z'-\text{factor} = 1 - \frac{3(\sigma_p + \sigma_n)}{|\mu_p - \mu_n|}$$

Where data follows a normal distribution and:
σ_p: standard deviation of positive control replicates
σ_n: standard deviation of negative control replicates
μ_p: mean of the positive control replicates
μ_n: mean of the negative control replicates

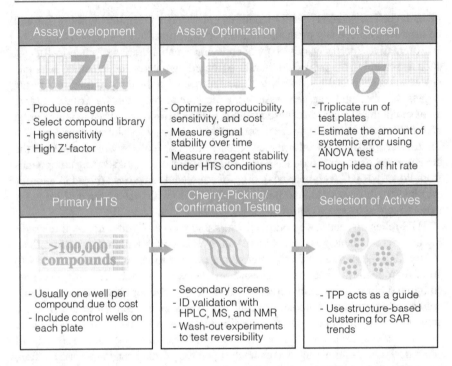

Fig. 2.4 General high-throughput screen workflow. Abbreviations: *ANOVA* Analysis of Variance, *HPLC* High-Performance Liquid Chromatography, *HTS* High-Throughput Screening, *ID* Identification, *MS* Mass Spectrometry, *NMR* Nuclear Magnetic Resonance, *SAR* Structure-Activity Relationship, *TPP* Target Product Profile

The typical HTS workflow can be broken into the following steps, which are illustrated in Fig. 2.4:

1. Procuring or scaling up production of the reagents (*e.g.*, proteins or cells, substrates, solvents, reporters)
2. Developing the assay, including miniaturization
3. Assay optimization (*e.g.*, Z′-factor, reproducibility, sensitivity)
4. Characterization of the optimized assay, such as sensitivity to time and temperature, linear range, and reagent stability
5. A pilot screen with triplicate runs of a small selection of the compound library
6. The primary HTS
7. Selection of actives and cherry-picking samples
8. Confirmation testing and counter-screens
9. Compound structure-based clustering
10. Confirmation of hits, evaluating the purity and identity of selected actives using liquid chromatography–mass spectrometry (LC-MS) and nuclear magnetic resonance (NMR), respectively, and confirming activity in secondary assays

HTS assays are typically run using 2–30 μl in 384 or 1536-well microtiter plates, although some assays are difficult to miniaturize beyond 96-well plates. The choice of assay technology often depends on the detection equipment available, cost of reagents (particularly for a screen of a large library of compounds), stability of the reagents, ease of use, and the potential for assay technology-dependent false positives.

Once we chose an assay technology, assay design and optimization involves tradeoffs between assay sensitivity to compounds, Z′-factor, and cost. If cost is not a consideration, you can often add large amounts of detection reagent and get both an enhanced Z′-factor and an increased sensitivity to inhibition by compounds. In practice, especially for an academic effort, cost is an important consideration. For enzyme assays, the choice of substrate concentration (relative to K_m) will affect the type of inhibitors or activators that are identified. The Michaelis constant K_m is the substrate concentration at which an enzymatic reaction rate is ½ the maximal reaction rate and is a measure of the affinity of an enzyme for its substrate. Running the assay with the starting substrate concentration equal to K_m will give the best overall sensitivity to competitive, uncompetitive, and noncompetitive inhibitors (Copeland 2003). Unlike most assays designed to study enzyme kinetics, HTS assays often allow substrate conversion to proceed to around 50%, since this produces much better signal/noise at a loss of only ~1.4-fold in sensitivity to competitive compound inhibition (Wu et al. 2003).

Phenotypic screens use a biological response (*e.g.*, cell death, protein translocation) to report compound activity. Because phenotypic responses reflect a complex biological cascade, they can be more accurate readouts of the therapeutic potential of a molecule if the assay accurately reflects a fundamental part of the disease process. Confirmation via secondary assays, however, can be more difficult, as the compound target in a phenotypic screen may be unknown.

Once the assay has been developed, it often requires optimization to obtain an adequately high Z′-factor and robustness. This is particularly true if the assay suffers from "edge-effects," a situation where the outside wells (in a control plate) have a bias toward different values than the rest of the plate. This can be caused by differences in temperature (plates warm up from the outside), evaporation, and in the case of plated cell-based assays, differential cell growth. It can take considerable experimental effort to identify the cause(s) of the artifact and redesign the assay to minimize its effects. As an example, for a thermal gradient edge effect, a long incubation with a lower enzyme concentration might replace a short incubation to allow time for thermal equilibration before assay readout.

Box 2.3: What Surprised an Academic?
Nearly all high-throughput screens identify reproducible (*i.e.*, not produced by variance) false positives. This is why it is so important to have secondary assays with different reporters to confirm hits. —*BK*

During assay optimization, its performance should be tested for sensitivity to varying conditions such as: (1) the concentration of the target protein (*e.g.*, for binding assays, does the amount of binding increase linearly with the amount of receptor added?); (2) the time of incubation of the assay (*e.g.*, for biding assays, has the assay reached equilibrium?); (3) time of preincubation of the individual reagents at room temperature (*i.e.*, do they have the stability required for many plate assay runs?); (4) solvent concentration (typically dimethyl sulfoxide, known as DMSO); and (5) sensitivity to standard compounds (if available). If it does not interfere with the assay, we should use high concentrations (*e.g.*, 5% v/v [volume per volume]) of DMSO, since this tends to increase the solubility of many compounds. However, cell-based assays are typically sensitive to DMSO, with the limit often being at 0.5–2% v/v.

After assay optimization, the assay protocol is "frozen" and a pilot screen is run to rigorously test whether the assay is ready for HTS. Three identical sets of compound plates (typically several thousand unique compounds) are run through the assay, one set (in randomized plate order) per run. The data are analyzed using analysis of variance to determine the sizes of the systematic errors due to plate order, plate row, plate column, *etc.* Ideally the variance is almost all "random," with only very small contributions from systematic errors.

> **Box 2.4: What Surprised an Academic?**
> I had assumed that converting our low-throughput phenotypic screen for anti-parasitic drugs to a high-throughput format would be fairly routine, but a number of difficulties arose when we began using 384 cell plates. We had to re-optimize cell density and had issues with strength of signal. Ultimately, we had to pivot to an imaging-based approach. The process required more time and work than I had expected. —*GE*

All HTS assay designs result in the identification of reproducible (*i.e.*, not produced by variance) false positives. These can result from compound interference with the assay readout or from undesirable modes of interaction with the target. Examples include reactivity of the test compound leading to covalent modification of the target, or compounds that inhibit the detection readout directly—a common concern in a luciferase-based assay. Thus, it is essential to develop additional independent assays to validate the hits (active compounds) from the primary screen or, if that is not possible, to eliminate potential mechanisms producing false positives.

These validation assays should seek to answer the following questions:

1. Does the compound interact directly and reversibly with the molecular target and with reasonable stoichiometry?
2. Does the reported structure of the active compound match what is in the well? Is it reasonably (>90%) pure? If the compound is <99% pure, is the activity quantitatively the same after purification or resynthesis?
3. Does the compound interfere directly with the reporter readout used?
4. Is compound activity quantitatively reproducible using a different assay technology (*e.g.*, cell-based versus in vitro)?

5. Is the activity reversible after washout? (The relationship between potency and expected off-rate should be considered.)
6. Is there evidence of a structure–activity relationship (SAR) for the active compounds? Are there related inactive compounds in the library?
7. Is the compound just generally reactive under the assay conditions? This can be assessed by comparing compound activity before and after incubation with potential target moieties or chemical groups (*e.g.*, 5 mM lysine dissolved in assay buffer).

Following these steps should result in a well-characterized primary screening assay and a set of secondary assays suitable for an HTS campaign in academia. Alternatives to activity-based HTS are listed in Table 2.1. These methods rely on the binding (measured or predicted) of a compound to its macromolecular target rather than disruption of the protein's activity. Except for virtual screening, they require natively folded purified protein.

Recommendations

- For a biochemical HTS assay, substrate concentration should be equal to K_m to help identify competitive, uncompetitive, and noncompetitive inhibitors. Different conditions may be required to identify activators (depending on the sensitivity of the assay). Usually, 50% of the substrate should be converted to product for optimal signal/noise.
- For cell-based assays, the percentage of organic solvents should be minimized, and solvent alone should be run as a control during assay development. Live cell imaging can be particularly challenging for large libraries unless the microscope system has a temperature, % CO_2, and humidity-controlled environment.

Resources

- PubChem: https://pubchem.ncbi.nlm.nih.gov/.
- Lilly/NCGC Assay Guidance Manual: https://www.ncbi.nlm.nih.gov/books/NBK53196/.
- Society for Laboratory Automation and Screening: https://www.slas.org/.
- SLAS Discovery Journal: https://www.journals.elsevier.com/slas-discovery.

2.4   Medicinal Chemistry and Lead Optimization

Daniel A. Erlanson

Lead optimization means taking a small molecule with promising properties and transforming this "hit" into a drug. It is like molecular sculpture, but instead of developing an aesthetically pleasing statue (which sometimes occurs), the aim is to

Table 2.1 Affinity-based high-throughput screening methods

Affinity-based HTS methods		
Library	*Method*	*Description*
Large, virtual libraries	Virtual screening	Virtual screening (Stumpfe and Bajorath 2020) utilizes computational models to screen very large libraries of potentially synthesizable molecules. In silico screening is not a complete replacement for real world assays; however, they may allow you to bypass HTS and proceed directly to Confirmation Testing
Large, pooled compound libraries	DNA-encoded compound libraries	DNA-encoded compound libraries (Gironda-Martínez et al. 2021) use DNA tags that are added sequentially during compound synthesis to produce very large pooled compound libraries. During several rounds of selection, the bound compounds are identified using PCR and DNA sequencing
	Affinity-selection mass spectrometry	In affinity-selection mass spectrometry (Prudent et al. 2021), mixtures of compounds are incubated with the macromolecular target, and then the target with any bound compounds is separated from the unbound compounds. The bound compounds are then freed by denaturation and identified by liquid chromatography coupled mass spectrometry (LC-MS)
Large and medium-sized plated libraries	Thermal shift assay	In a Thermal Shift Assay (Gradl et al. 2021), the binding energy of a compound leads to a higher melting temperature for the protein, which can be assayed in a thermal cycler using SYPRO Orange, a dye that selectively binds to unfolded protein
Medium-sized plated (*i.e.*, one compound/well) libraries	SPR High-throughput crystallography	Biophysical methods for screening medium-sized plated (*i.e.*, one compound per well) libraries are often used to screen lower molecular weight compounds called "fragments" (Murray and Rees 2009). Surface Plasmon resonance spectroscopy and high-throughput protein crystallography are popular methods

Abbreviations: *HTS* High-Throughput Screening, *SPR* Surface Plasmon Resonance

construct a safe and effective molecule for treating a specific disease. And instead of chisels and plaster, practitioners—medicinal chemists—apply the tools of chemical synthesis.

The previous section covered high-throughput screening (HTS) which, if successful, has generated a hit—a small molecule that has some activity for the target or phenotype of interest. Of course, this hit is likely a long way from being a drug. Improving affinity is often the first task of lead optimization. A drug should be as potent as possible to reduce the cost of production, minimize the size of the pill or injection needed, and reduce the potential for off-target effects. Most drugs have IC_{50} or EC_{50} values (half maximal inhibitory concentration or half maximal effective concentration) around 10 nM or so, with considerable variation to either side. Hits from HTS are sometimes nanomolar potency, but more often low micromolar, which means that the binding affinity may need to be improved by several orders of magnitude. Strategies for lead optimization are illustrated in Fig. 2.5 and are further discussed in this section.

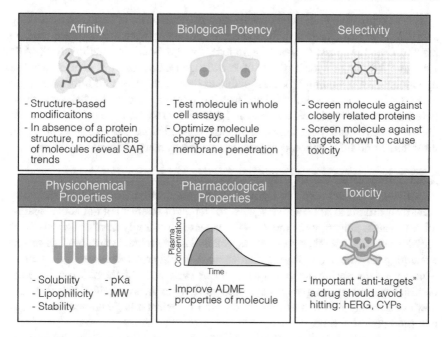

Fig. 2.5 Strategies for optimizing a lead compound. Abbreviations: *ADME* Absorption, Distribution, Metabolism, and Excretion, *CYP* Cytochrome, *hERG* Human Ether-à-Go-Go-Related Gene, *SAR* Structure-Activity Relationship

2.4.1 Lead Optimization Considerations

2.4.1.1 Improved Affinity

Knowing how the molecule binds can generate ideas to improve potency. For example, there may be a pocket on the protein near the small molecule binding site, and adding a chemical group (or moiety) to reach this pocket may pick up additional interactions and thus additional binding energy. Alternatively, a structure may reveal an unfavorable contact: Perhaps a hydrophobic (water-hating) portion of the ligand is exposed to solvent, or a hydrophilic (water-loving) portion is buried in a greasy hydrophobic part of the protein; the medicinal chemist would make analogs of the molecule without the unfavorable contact and test the activities of the new molecules.

Ideally this will lead to better potency, but often changes are less dramatic than expected, and additional molecules will need to be made. This iterative process is called structure-based drug design. In the best cases, it is possible to obtain structural information regarding how the small molecule binds to the target using experimental techniques such as X-ray crystallography or NMR spectrometry. Failing this, computational modeling can give some idea of the binding mode if the structure of the target is known or is believed to be similar to another characterized target.

It is also possible to do lead optimization in the absence of structure by making somewhat random changes to the molecule and seeing what effects these have on activity. Over the course of several iterations, structure–activity relationships (SAR) emerge. SAR can provide a wealth of knowledge that a medicinal chemist can use to understand the binding mode. Although experimental structural information has become a key tool in medicinal chemistry, it is worth remembering that X-ray crystallography was not sufficiently rapid and general for routine use until the 1980s and 1990s, and even today medicinal chemistry is applied to many targets for which direct structural information is not available, such as many membrane proteins.

2.4.1.2 Improved Selectivity

Selectivity is another critical factor in lead optimization. Researchers generally want their drug lead to be active against the target of interest but not active against other proteins. Selectivity is most readily assessed by simply measuring activity of the molecule against other proteins, especially closely related ones, but this can be a daunting task. For example, there are more than 500 protein kinases in the human genome, so measuring activity against all or even most of them can get pricey. Fortunately, enough companies have been working in the kinase field that there are now commercial offerings to confirm selectivity against a large number of kinases in a short period of time.

However, such selectivity testing for newer classes of targets and enzymes is often not available. Note that selectivity testing within a related family of enzymes or receptors does not rule out the possibility that a compound will bind to a protein outside that family. Before compounds advance into the clinic, they are tested against a panel of up to several hundred targets that could cause problems (see below). However, not everything can be tested in vitro, and off-target effects often manifest as side effects and toxicity during in vivo studies.

2.4.1.3 Improved Physicochemical Properties

Throughout the course of lead optimization, it is important to keep an eye on the physicochemical properties of the molecule such as solubility and lipophilicity (the way it partitions between water and oil or membranes). Solubility can be a tricky balancing act because improving potency often involves increasing the size and lipophilicity of a molecule, leading to decreased solubility. Chemists sometimes refer to particularly insoluble compounds as "brick dust."

2.4.1.4 Improved Biological Potency

Initial screens are often conducted using pure isolated proteins under highly artificial conditions. Therefore, it is essential that potency be determined in more biologically relevant systems such as whole cell assays; often compounds that show activity against the isolated protein will show less or no activity in cells. Sometimes this is due to factors that a medicinal chemist may be able to fix rationally. For example, compounds that are negatively charged can have difficulty crossing cell membranes to interact with targets inside the cell. In other cases, it is unclear why there is a disconnect; in these cases, it may be necessary to make more dramatic changes to the lead series, or switch to another series entirely.

2.4.1.5 Improved Pharmacological Properties

Potency and selectivity are important, but other parameters also require optimization. In fact, a rookie mistake is to focus exclusively on potency. Many things can happen to a drug on its way to its target. This is especially true for oral drugs: The body treats anything coming in through the mouth as food and tries to digest it or, failing that, to excrete it. The study of what happens to a drug in vivo is called pharmacokinetics (PK), which is covered in more detail in Sect. 3.4. A critical aspect of lead optimization is to measure and improve the ADME properties of a molecule, keeping it in the body for long enough and at sufficient levels to do its job without causing problems. Many of the individual proteins that affect a drug's path into and through the body are known, and experiments with isolated enzymes, plasma, or liver extracts can be helpful, but ultimately animal studies are essential to truly understand a molecule's PK.

Because so many different factors are at play in pharmacokinetics, medicinal chemists often turn to empirically derived rules to try to tune the properties of their molecules. The most famous of these is Chris Lipinski's Rule of 5, a set of guidelines concerning molecular weight, lipophilicity, and other properties that predict the likelihood a drug candidate will be orally bioavailable (Lipinski et al. 1997). When performing SAR to optimize PK, often a specific moiety of the molecule may be prone to metabolism, and by altering this bit of the small molecule, the overall stability can be improved. Keep in mind that such rules are not hard cutoffs, but directional guidelines to improve the probability of success.

2.4.1.6 Target Validation

Pharmacokinetics is sometimes characterized as "what your body does to a drug." Conversely, pharmacodynamics (PD) can be thought of as "what a drug does to your body." On a fundamental level, the drug needs to be active against the target of interest.

Unfortunately, it is possible to inhibit or activate a biological target and yet have no effect on the disease of interest—this is particularly true for newer targets. Validated targets are targets for which modulation of their activity alters a disease state, and the best way to validate a target is by using a small molecule (or peptide or protein). A tool compound can be used for target validation—this is a molecule that has sufficient activity and ADME properties to answer basic biological questions about the target, but may not be suitable as a drug, perhaps because it is toxic or has other deleterious properties. For more on target validation, see Sect. 2.1.

2.4.1.7 Reduced Toxicity and Drug–Drug Interactions

There is a growing consensus that virtually all drugs have off-target effects, and it is important to understand these and determine whether they will cause adverse events. Toxicology is concerned with specific toxic effects, for example, liver damage. A number of molecular substructures are known to have caused toxicity in the past, and medicinal chemists try to avoid having these moieties in their lead molecules. Ultimately though, it is impossible to predict whether a given molecule will be non-toxic without doing in vivo experiments.

Moreover, toxicity is not the only problem; a drug lead should avoid hitting many other "anti-targets." One of the most important is a cardiac ion channel protein called hERG, which when inhibited can cause severe and sometimes fatal heart problems. This has led to the withdrawal of several marketed drugs, and medicinal chemists today almost universally assess the hERG activity of their leads. The SAR of hERG binding is partially understood, and often medicinal chemists can engineer promising leads to maintain potency against the target protein while also avoiding hERG binding.

Similarly, many enzymes involved in metabolizing drugs (particularly a large class of enzymes called CYPs) can be inhibited by small molecules, leading to drug-drug interactions if the enzymes in question are necessary for metabolizing other drugs. During lead discovery, it is important to measure CYP inhibition and, ideally, make changes to the molecule to reduce or eliminate it.

Pharmacokinetics and pharmacology are both dependent on animal models, but it is important to always remember that mice are not furry little people: Drugs metabolized rapidly in mice may be stable in humans and vice versa. Because of such differences, obtaining animal data in at least two different species is usually necessary before moving a drug into the clinic.

Box 2.5: What Surprised an Academic?
We started KAI with three drug candidates for three different clinical indications. When asked by the venture capitalists (VC) to rename the drugs (to differentiate them from those used in my academic laboratory), I thought it was silly that they did not accept the names KAI 001, KAI 002 and KAI 003. In my naiveté, I was sure that we would not need to make more than 999 compounds after all the preliminary work in my university lab. I also did not realize that a company should not reveal to others how many compounds were made (*e.g.*, if few were made, the IP might not be as strong). So instead of giving sequential numbers, our VC dubbed KAI-9803 based on my answers to "what year did you design that peptide?" and "where did it fall in the sequence of peptides you designed that year?" —*DM-R*

2.4.2   Other Issues

A recent trend in medicinal chemistry is fragment-based drug discovery. Instead of starting with low micromolar IC_{50} lead-sized or drug-sized molecules, this approach starts with smaller "fragments" with molecular weights one-quarter to one-half the size of typical drugs and potencies in the mid to high micromolar range. Because there are fewer small fragments than larger molecules (just as there are fewer two letter words than four letter words), it is possible to screen chemical diversity more efficiently. Moreover, smaller, simpler molecules are less likely to have extraneous bits that do not help the overall potency but cause problems with PK or PD. Of course, identifying and optimizing lower affinity molecules present challenges in their own right.

Ultimately, lead optimization requires the medicinal chemist to improve numerous parameters simultaneously: potency, selectivity, solubility, PK, and PD. Unfortunately, improving one may exacerbate another. Medicinal chemistry requires picking the best possibilities to explore, even though it is impossible to gather all data for every compound.

In fact, there is no guarantee that it is even possible to produce a molecule that satisfies all the necessary parameters; targets for which this happens are called "undruggable." This multiparameter optimization in the absence of complete data is what makes medicinal chemistry as much an art as a science, and the fact that a solution may not exist sometimes makes it a particularly frustrating art. The next time you take a drug, it is worth reflecting on the effort, skill, and serendipity that went into discovering that little molecular sculpture.

Box 2.6: What Surprised an Academic?
When our high-throughput screen to correct a mutant enzyme identified three halogen analogs of a small molecule that increased the activity tenfold, we were sure that we had found our lead compound for clinical trial. After hearing from VCs that we should identify superior analogs, we secured funding to conduct further medicinal chemistry. You can imagine our frustration when our effort failed. We realized that there is much 'art' to medicinal chemistry when a company that licensed our technology managed in a very short time to identify a compound with superior efficacy and more desirable pharmacology. —*DM-R*

The Bottom Line
Lead optimization to obtain a drug candidate requires the chemist to optimize many parameters simultaneously including potency, selectivity, solubility, lipophilicity, protein binding, tissue distribution, PK, and PD. Medicinal chemistry requires selecting the best structural modifications to explore, often with incomplete data and without any guarantee that an acceptable solution even exists. Medicinal chemists frequently rely on their intuition, making drug optimization both a science and an art.

Resources
- Journal of Medicinal Chemistry. This is probably the premier journal for medicinal chemistry but has onerous requirements for compound characterization. http://pubs.acs.org/journal/jmcmar.
- Nature Reviews Drug Discovery. This is a good source of information on all aspects of drug discovery. https://www.nature.com/nrd/.
- In the Pipeline. This is probably the best chemistry-related blog out there. The author, Derek Lowe, is an experienced medicinal chemist who writes prolifically about a range of topics, and his posts attract dozens of comments. https://www.science.org/blogs/pipeline.
- Practical Fragments. For all things dealing with fragment-based drug discovery and early-stage lead optimization, my blog is a good resource. http://practicalfragments.blogspot.com/.

2.5   Natural Products

Julie Saiki

2.5.1   Drugs Developed from Natural Products

Natural products, which are small organic molecules produced by any living organism found in nature, have been used to treat a wide range of human diseases for thousands of years (Ji et al. 2009). Approximately 50% of currently available FDA-approved drugs have been developed from natural products derived from plants, animals, or microorganisms. Among them are the antibiotic penicillin, cancer therapy paclitaxel, immunosuppressant cyclosporine, lipid-lowering lovastatin, and the antimalarial drug artemisinin for which Chinese scientist Tu Youyou won the 2015 Nobel Prize in Physiology or Medicine (Cragg and Newman 2013).

Botanicals are an important subset of natural products that are derived from plants, algae, fungi, or their combinations. Some botanicals, such as herbal or dietary supplements, may be valued for their perceived health benefits, but have not been proven to treat, cure, or prevent a medical condition and cannot be marketed to make such claims in the US. FDA regulates a botanical drug in the same way as any other prescription pharmaceutical drug: It must go through a rigorous process to evaluate its safety and efficacy before it can be sold with a claim to prevent or treat a disease (U.S. Food and Drug Administration. Center for Drug Evaluation and Research (CDER). Botanical Drug Development: Guidance for Industry 2016).

One challenge with botanical drug development is that botanicals are naturally heterogeneous mixtures. The therapeutic effect may be due to one or a mixture of structurally diverse chemical components (Li and Vederas 2009). Their biological origins may provide an advantage since these compounds were evolutionarily selected to bind to proteins such as enzymes, protein receptors, and transporters. As a result, they may be more adept at binding to protein targets that are important for the treatment of disease (Li and Vederas 2009). Botanical drugs, especially those developed from traditional Chinese medicine, also have historical traditions that encompass thousands of years of observational experience and may have a higher likelihood of having biological activity and favorable safety profiles. Although the pharmaceutical industry has focused modern drug discovery efforts on screening synthetic libraries, natural products continue to be an important source for new drug discovery.

2.5.2   How to Develop Drugs from Botanicals

Identifying new drugs from botanicals has unique challenges. Because a defined chemical composition may not be known, complex mixtures have the potential for multiple active ingredients acting synergistically. More complicated are formulations that combine several natural products (*e.g.*, plant extracts) with the belief that some ingredients increase efficacy, whereas others mitigate toxicity (Hsiao and Liu 2010). Modern instrumental techniques (such as spectroscopy and chromatography) can be used to characterize complex mixtures and identify single molecules. However, reducing a mixture of compounds to a single molecule can in some cases result in loss of efficacy or safety. If a single active ingredient can be identified, medicinal chemists may be able to synthesize the naturally occurring compound and create novel patentable derivatives, analogs, or prodrugs with improved efficacy, safety, pharmacokinetic, or pharmacodynamic properties. However, because of the unique chemical structures of natural compounds, it is not always possible to synthesize a compound de novo, and the crude purified natural product may be needed for the starting material.

FDA has issued a Guidance for Botanical Drugs designed to enable the development of botanical products as drugs (U.S. Food and Drug Administration. Center for Drug Evaluation and Research (CDER). Botanical Drug Development: Guidance for Industry 2016). This guidance outlines two key requirements for an IND that differ from traditional synthetic drugs. The first requirement is to define a chemical "fingerprint" that characterizes the product and ensures batch-to-batch consistency, since botanicals are often complex mixtures lacking a single active ingredient. The second is to provide available evidence of prior human use, since many botanicals have historical experience. Some botanicals may even have been previously marketed or tested in clinical studies and reported in the literature. This information is critical, as prior human experience may modify the amount of nonclinical information required to support initial human studies. Key additional features of the FDA guidance for botanical drug development are:

- A botanical product may be classified as a food (including a dietary supplement), drug (including a biological drug), medical device, or cosmetic. Whether an article is a food, drug, medical device, or cosmetic depends in large part on its intended use, though for some product types, other factors must also be considered.
- The clinical evaluation of botanical drugs in early-phase clinical studies is robust and does not differ significantly from that of synthetic or highly purified drugs in such studies.
- Pre-IND, end-of-phase 1, end-of-phase 2 and 2A, prephase 3, and pre-New Drug Application (NDA) consultations with FDA are strongly encouraged to assess the adequacy of existing information for submission of an IND or NDA, obtain advice regarding the need for additional studies, ensure that clinical protocols are properly designed, and allow discussion of the initial or overall development plan.

Despite hundreds of botanical product IND applications submitted to FDA for clinical investigation in a wide range of diseases (Wu et al. 2020), only two botanical drugs are commercially available: (1) Veregen (sinecatechins), a green tea extract for topical treatment of external genital and perianal warts (*Condylomata acuminata*) in immunocompetent adults, and (2) Mytesi (crofelemer), an oral extract from the latex of bark from the South American *Croton lechleri* tree that is indicated for symptomatic relief of noninfectious diarrhea in adults with HIV/AIDS (human immunodeficiency virus/acquired immunodeficiency syndrome) receiving antiretroviral therapy. Challenges in manufacturing and lack of patentability have likely hampered development of botanicals in industry. Nonetheless, botanical drug development remains a viable and important path for novel treatment of diseases with unmet medical needs (Wu et al. 2020; Ahn 2017).

Box 2.7: What Surprised an Academic?
The traditional drug regulatory path requires a standard battery of costly and time-consuming GLP preclinical safety studies before submitting an IND application. If there is adequate documentation of prior human experience, the botanical drug regulatory path may allow you to skip many of these studies, thereby significantly accelerating the timeframe and reducing costs to get to an early-stage clinical study. When we were developing a botanical extract that is extensively used in traditional Chinese medicine to treat patients with treatment-refractory ulcerative colitis, we were surprised to learn that we could use safety data reported in the literature regarding prior human use and bypass preclinical safety studies. After establishing the chemical fingerprint of the extract and characterizing the heavy metal and microbial impurities, we were able to submit an IND and advance rapidly to a phase 1b clinical study in patients. —*JS*

The Bottom Line
Natural products are a rich source of bioactive compounds that have therapeutic potential. Plant extracts have been extensively used as traditional medicines over the past two millennia. FDA has issued a Guidance for Botanical Drugs that may allow evidence from prior human experience to expedite the path to a clinical trial.

2.6   Approaching Machine Learning in Drug Development

Jennifer L. Wilson

Machine learning has a revived reputation in biological sciences, although this is not the first debut of the technology. Forms of machine learning such as regression have been around for decades, whereas neural network architectures have recently matured with wider access to increased computing power. There are common aspects in all of these methods—the algorithms find patterns among the data and use these patterns to make new predictions or classifications. Successful application of these methods depends on many features. If applied correctly, machine learning can increase the scale and impact of drug discovery. For instance, a well-designed machine learning analysis can identify new uses for existing drugs, recognize patterns for rare events of drug toxicity, prioritize important cellular pathways for diseases and side effects, and elucidate molecular structures with favorable target binding.

2.6.1 Selecting Data Makes a Useful Model

An important component of effectively developing and deploying these models is finding the right data to model. It is possible to develop a highly performant model using a broad range of data, but that model might not be useful if the input data are not sufficiently connected to the biology or if model predictions are not connected to outcomes, such as improved patient care (Char et al. 2018). A famous example used word embeddings—reduced representations of spoken and written language—to assess an algorithm's ability to solve analogies (a task similar to the infamous SAT problem). The algorithm performed well, yet also discovered societal biases when asked to solve analogies including: "King:Man, Queen: ___, and Doctor:Man, Nurse: ___". In both of these scenarios, the algorithm answered "woman." The first makes semantic sense, but the latter represents a bias in the underlying data. Women and men can be nurses, but the algorithm favored selecting "woman" over "man" because the word "nurse" was more often associated with "woman" in spoken and written language. Machine learning further captured historical perspectives and shifts regarding stereotypes and ethnic minorities (Garg et al. 2018). Increasingly, the biomedical research community has acknowledged similar biases in research data; specifically, many populations are missing from popular data resources, especially genetic data repositories. Therefore, it is especially important to understand representation in the underlying data before developing a model for a given population.

2.6.2 Machine Learning Predictions Can Change with Time

Predictive algorithms have the power to amplify patterns in the data, and interestingly, this characteristic can either be a "bug or a feature." First is an example of a feature: When identifying gene sequences to generate novel antimicrobial peptides, a combined generative network and predictive algorithm created a virtual, directed evolution pipeline. The team generated gene sequences with some noise, used a discriminator algorithm to predict the likelihood of antimicrobial properties, and then used the highest scoring sequences as training data for a new generative algorithm. They repeated the process with new input sequences to eventually converge on a discriminator algorithm that was increasingly sensitive to sequences with the desired properties. In this case, amplifying patterns in the input data "evolved" better sequences for the development of antimicrobial peptides (Das et al. 2021).

However, amplification of patterns within data can also have negative and undesirable consequences—"a bug." In medical decision-making, algorithms are routinely used to make decisions about patient care. These "algorithms" include simple methods, such as rule sets or more complicated neural networks, applied to patient electronic health records (EHR). Ethical uses of machine learning in medicine should be considered. An example is the allocation of scarce resources, such as transplant organs. In this scenario, physicians may consider a patient's overall health and other underlying conditions before allocating an organ. A machine learning algorithm may "learn" these preferences and infer that higher overall health is required for organ transplant success. An algorithm deployed in this space could amplify patterns in care, and as the algorithm is deployed in healthcare settings, it could continuously bias physicians to allocate organs to the same type of patient (Char et al. 2018).

These two examples highlight the importance of understanding whether your data are dynamic or static. Static data are highly amenable to machine learning, whereas dynamic data could introduce nonintuitive outcomes if temporal changes are ignored. In the peptide example, the antimicrobial molecular qualities are fixed even though the authors updated their data to converge on the most useful sets of peptides. In the transplant organ case, decisions about allocating resources are dynamic; a patient's health status as well as the number of available organs may influence whether a patient receives a transplant. Keep in mind that the medical environment is dynamic and influenced by multiple factors.

2.6.3   Use Machine Learning to Replace Well-Defined Tasks

Machine learning algorithms can be incredibly useful in cases in which they abstract and replace tedious analysis tasks. Some of the earliest applications of machine learning were in labeling images. Familiar examples include labeling images as "cats" or "dogs," or medical applications for classifying tumor images as malignant or benign. Image-based classification can also be useful in high-throughput imaging-based screens for tasks such as labeling cells or organelles or classifying cellular morphology (Godinez et al. 2017). Pharmacokinetic parameters, including maximum concentration, time to maximum concentration, area under the concentration curve, and drug half-life, are important for advancing therapies with desired dose-response relationships. In cases where dose-response curves are not easily fit with linear or polynomial equations, some have proposed using machine learning to directly extract these parameters from concentration vs. time curves.

Machine learning methods, especially deep learning, can also reduce the necessity to preprocess data. For instance, a survey of published predictive algorithms discovered that most implementations used a median of only 27 predictor variables to assess risk, despite the number of immense data variables available in EHR (Goldstein et al. 2017). In fact, deep learning methods can accommodate the richness and scale of EHR data and can make predictions without extensive feature engineering, eliminating the need to reduce input features to single digits (Rajkomar et al. 2018). The authors of this last study further demonstrated an ability to extrapolate prediction to multiple healthcare centers; often models trained on one healthcare system are not portable to other healthcare systems due to features innate to each healthcare system. In the latter example, it was not required to clean and merge data from disparate healthcare systems before beginning predictive modeling. Machine learning algorithms may employ various forms of regularization to penalize or reduce the number of features and can simultaneously prune datasets during learning.

2.6.4   Avoid the Black Box If You Need to Understand the Biology

As well as predictive tasks, machine learning methods can also be used for attribution or advancing understanding of biological systems. For predictive tasks, model performance on novel data is of high priority and "black box" models may be used. Models such as neural networks and deep learning can be highly performant and highly accurate; however, it is difficult for humans to interpret how the model is making a particular prediction due to high dimensions and nonlinearity. Furthermore, the relative importance of outputs and how the outputs interrelate with each other is unclear. In contrast, models such as linear regression or decision trees can more easily be dissected to understand which data features are used for prediction. These models are considered "white box" because it is possible to relate prediction back to data features. Model performance can still be high if the input data have strong relationships to the predicted output.

The following are two examples of "white box" machine learning. In the first example, we conducted a pathways analysis of multiple drugs with similar severe side effects. We used all network proteins as input to a logistic regression model and assessed prediction performance to ensure that there was a relationship between network proteins and drug effects. We next used regression parameters to infer network proteins with high predictive values and repeated this process for a set of severe side effects. Interestingly, proteins downstream of drug targets had high regression weights, suggesting that drug targets are not solely responsible for drug-induced effects. Using a unique model for several side effects, we identified highly predictive proteins unique to each side effect and hypothesized novel mechanisms for drug-induced effects (Wilson et al. 2022). In this example, we tried to attribute the side effect to the downstream proteins.

In the second example, we used parallel machine learning algorithms—specifically, two graph convolutional neural networks—to first learn features associated with protein binding pockets and to learn chemical features of ligands (Torng and Altman 2019). After training these two models, they predicted ligand-protein interactions without prior crystal structures of ligand-protein binding. Within their model, the algorithms extracted parameters for combinations of amino acids in protein pockets and atoms in the ligand chemical structures. Using these parameters, amino acid to ligand structure interactions could be ranked to better understand why the algorithm was predicting ligand-protein interactions. In a specific example, the algorithms characterized binding between the tyrosine-protein kinase SRC (SRC) and pyrazolopyrimidine-5 and discovered that valine 323 in the SRC pocket and the nitrogen and oxygen atoms in pyrazolopyrimidine-5 had high importance for binding prediction. Investigating algorithm parameters supported attribution of prediction to individual amino acid-atom interactions and led to the development of testable, biological hypotheses (Torng and Altman 2019).

Another intriguing integration of machine learning with prior knowledge used high-throughput perturbation datasets in combination with ordinary differential equations of cellular pathways. Typically, rate equations for ordinary differential equations are meticulously tested using cellular assays or are discovered through literature searches. Instead, the authors incorporated ordinary differential equations (ODEs) with TensorFlow to directly infer quantitative pathway parameters. By using machine learning to infer ODE interaction parameters, they generated an interpretable model of cellular responses. The researchers demonstrated model performance to predict melanoma cell response to several therapies (Yuan et al. 2021). These integrated approaches demonstrate a new power of machine learning applications—the ability to integrate large amounts of data while still providing insights about "how" the algorithm made decisions.

Box 2.8: What Surprised an Academic?

When a computer scientist colleague asked for assistance in validating her human proteomic based machine learning model to predict repurposed drugs for any clinical indication, I provided a list of 36 diseases from our SPARK projects. In each case, we had evidence of efficacy for a repurposed drug in either animal or human studies. To my surprise, in 34 of the 36 cases, the model identified our drug among the top 15. In fact, the two indications that were missed were both viral infections and our drug targeted a viral protein—truly amazing predictive performance! In most cases, however, the algorithm ranked our drug after 9 or more other drugs. Would those drugs also be efficacious? Since our colleague was using a "black box" approach, we had no way to prioritize the additional possible repurposing candidates. At a cost of ~$100,000 per drug to provide preclinical proof of efficacy, we did not have the resources to explore these alternate candidates. If we had some way to vet the predictions (*i.e.,* a "white box approach"), we may have been able to identify high priority candidates for further study. —*KVG*

2.6.5 Design the Experiments with the Model in Mind

Machine learning applied to drug discovery can be powerful when the experiments and the modeling are both considered at initial study design. Consider a new opportunity to use machine learning to predict a new and efficacious drug for blocking phagocytosis. You have a new imaging-based assay with a fluorescent readout that maps to cellular viability and organelle shape. You deploy a large drug screen to test new drugs in your assay, including well-established blockers of clathrin as controls. You utilize a machine-learning approach to analyze images from the screen results and predict new blockers of phagocytosis. Using this approach, your algorithm would "learn" how the clathrin-inhibitor drugs alter phagocytosis and you'd only discover drugs that prevent clathrin pit formation. Phagocytosis is a complex biological process, however, that can be blocked at multiple steps leading to other changes in cellular morphology. To predict compounds that altered phagocytosis more broadly, you would have to include other control conditions that, for instance, prevent lysosomal fusion, actin-myosin contraction, or phagolysosome acidification.

Box 2.9: How to Move Forward

As you consider whether to apply machine learning to your project, we encourage you to think about a series of priorities:

1. Most importantly, what data are available to solve your problem? Do you have a solid biological rationale for using these data?

2. What do you hope to learn from your algorithm? Are you looking for high prediction performance or do you need to attribute output data to certain input features? Black box algorithms are useful for purely prediction tasks, but they are limited for understanding how a biological system works.

3. What computing resources are available to you? Simpler machine learning algorithms such as Logistic Regression can be run on a standard laptop. However, neural networks will need dedicated computational resources.

Other tips for success:

- Measuring predictive performance can be useful to understand if a relationship between an input dataset and output data set exists. If there is a low prediction performance, you might need an additional dataset or a reframing of the overall problem.
- Most machine learning approaches will require some amount of "tuning". Splitting data into training, testing, and validation sets is standard practice to both understand the utility of the model and prevent overfitting.
- Be skeptical of highly performant models—an extremely high model performance can be indicative of overfitting and you might be recovering a biological relationship that is already well known.
- Like other skills in this book, establishing a team is the best way to be successful.

The Bottom Line

Many successful machine learning models are black box, but it may not be clear how or why a model is making a particular prediction. Other models are "white box" or interpretable and may be desirable if you need to understand how the model is prioritizing input data to make a prediction. Choose white or black box models depending on your use case.

2.7 Therapeutic Antibody Discovery

Shelley Force Aldred

Natural immune systems have evolved to create their own therapeutic molecules, including a tremendously diverse collection of antibodies that bind with high specificity and affinity to a wide variety of targets. Leveraging lessons from nature, researchers can now create fully human antibodies in the lab to be used as drugs. Antibodies provide therapeutic action through mechanisms like ligand blockade, activation, or inhibition of receptors, induction of apoptosis, targeted toxin or conjugate delivery, and by Fc-mediated stimulation of ADCC (antibody-dependent cellular cytotoxicity), ADCP (antibody-dependent cellular phagocytosis), or CDC (complement-dependent cytotoxicity). While antibodies are typically monospecific (bind one target), their inherently modular structure enables engineering of bi- or multispecific molecules that engage more than one target at a time, either on the same cell or bridging two cells (Smith 2015).

Therapeutic antibody discovery and development activity has increased tremendously in recent years, with FDA approving the 100th monoclonal antibody (mAb) product in 2021 (Mullard 2021). As with any type of drug modality, therapeutic antibodies come with their own lists of pros and cons. Advantages for antibodies include exquisite specificity, long half-life, and low immunogenicity risk if they are fully human sequences. In terms of disadvantages, antibodies are limited to targets that are circulating ligands or cell-surface targets, and as large molecules they have challenges with tissue penetration and crossing the blood brain barrier. Compared to small molecules, they have more limited delivery mechanisms and higher production costs (Smith 2015; Ho 2018).

A full review of therapeutic antibody discovery and development warrants its own textbook, so this section provides a high-level overview of challenges and strategies. Specifically, a wide variety of de-risking strategies are presented, because the risk of drug candidates dropping out at each stage of development is high. While the breadth of the lists can be overwhelming, researchers may not need to check every one of these boxes. Keep in mind that every extra piece of data added to a preclinical package will de-risk ideas and molecules, increasing the value of the program and the likelihood of finding an industry partner.

2.7.1 Layers of Risk

2.7.1.1 Concept Risks

Which biological target(s) will be engaged, with what type of antibody-like molecule and with what desired mechanism of action? These three key components of a therapeutic concept represent unique axes of risk in the eyes of a potential partner, with most partners willing to tolerate risk on one or two of the axes. For example, if the concept includes engagement of a novel biological target, partners will be more likely to consider licensing if the candidate molecules rely on a well-understood mechanism of action (like ADCC) and a format that looks much like a natural antibody.

1. *Biological target(s)*: Proposing to engage a novel or not-yet-well understood target antigen presents a unique set of challenges. First, researchers must convince a partner that an antibody can be used to achieve the desired biological effect (biological proof-of-concept). The preclinical proof-of-concept burden will be significant, likely involving multiple in vitro and in vivo model systems. Second, researchers will need to show that the target is relevant for patients in the clinic and the market is big enough to warrant an expensive development program. For example, if the biological target is a newly discovered tumor associated antigen, you must be prepared to answer questions like: (i) What kinds of tumors express the antigen and what percent of tumors in each indication are positive for expression? (ii) Is the antigen expressed on healthy tissue anywhere in the body? (iii) What is the range of antigen density observed across patient samples and compared to expression on normal tissues?
2. *Mechanism of action (MOA)*: If the concept relies on a novel or relatively new MOA like blocking a newly discovered ligand/receptor interaction or using a bispecific antibody to bridge two cell types in a new way, the likelihood of finding an industry partner will increase if the biological target(s) are well known and the molecule looks like a natural antibody (like an IgG).
3. *Format*: Natural antibodies have evolved to have good therapeutic properties, and their modular nature makes it possible to add, remove, combine, or mutate various components while often maintaining apparently acceptable structure and function. However, each step away from natural antibody sequences or structures presents new manufacturing and immunogenicity risks. Therefore, preferred molecule designs will only include variations from natural characteristics that are absolutely required to achieve a specific biological effect. Where variations are called for, prioritize alternative structures or sequences that have been validated in other development programs. For example, many therapeutic antibody programs require a "silent" Fc to avoid engagement of key Fc receptors, but researchers have successfully advanced a variety of variant Fcs through manufacturing and into the clinic, providing a robust starting point for new molecules (Bates and Power 2019).

2.7.1.2 Candidate Molecule Risks

Creating or identifying an antibody or antibody-like molecule with a desirable functional profile is critical, but a strong lead must also have favorable drug-like properties. It is risky to assume that any single antibody with desirable function can be turned into a developable candidate, and few partners are willing to shoulder all the risk. The suite of characteristics often called "developability" include feasibility of manufacture, stability during storage, and absence of off-target stickiness. A variety of preclinical assays have been developed to help predict some of these characteristics, but such assays are not perfect predictors of downstream behavior (Jain et al. 2017).

Predicting clinical toxicity and safety profiles can be challenging and is covered by another section in this book (Sect. 3.7), but all therapeutic antibody programs will benefit from reducing immunogenicity risks. The first therapeutic antibodies were murine or chimeric human/murine molecules and were immunogenic, meaning recognized as foreign when injected into patients, prompting significant concerns for both safety and efficacy. Therefore, partners strongly prefer fully human antibodies, either discovered as fully human sequences or humanized from wild-type or chimeric animal sequences. Sequence changes introduced during a humanization process often alter the functional or developability profile of a lead antibody, and the original characteristics cannot always be rescued (Chames et al. 2009).

> **Box 2.10: What Surprised an Academic?**
> Many researchers assume that almost any antibody sequence that provides the appropriate functional activity can be turned into a therapeutic candidate. This is rarely the case! —*SFA*

2.7.2 Establishing Biological Proof-of-Concept

For any new therapeutic antibody development program, it is critical to show that an antibody can be used to achieve the desired biological effect, like demonstrating that an antibody to block a specific ligand/receptor interaction can drive tumor regression. Developing a lead panel of fully human and developable drug candidates is costly and time consuming, so such biological proof-of-concept evidence is an important gating item for many programs. In many cases, researchers can complete this work using easily accessible commercial tool antibodies or by creating new antibodies with a low-cost discovery platform like immunizing a wild-type animal. It is possible to convert a tool antibody into a drug candidate, but technical success is far from guaranteed and intellectual property issues must be considered.

2.7.3 Lead Discovery and Early Prioritization

With biological proof-of-concept in hand, researchers often shift attention to generating panels of novel candidate therapeutic antibody molecules. There are many factors to consider in discovery and screening designs; this is a resource-intensive phase of work. In addition, criteria laid out in a Target Product Profile (TPP) may be better addressed using one platform or strategy over another, so it pays to start with the end in mind (see Sect. 1.4 on Target Product Profiles).

2.7.3.1 Discovery Platforms
At a high level, there are two types of antibody discovery strategies: (1) isolating antibodies from a natural immune system, either from a human sample or an immunized animal (our focus here) or (2) selecting antigen binders from an existing pool

of antibodies with in vitro display platforms. Each approach has specific advantages and disadvantages (Laustsen et al. 2021).

Isolating antibodies from immunized animals is appealing because a natural immune system generates antigen-specific antibodies that are affinity matured endogenously (within that animal), often favoring antibodies with robust expression. However, achieving species cross-reactivity and humanizing the antibodies is technically and financially challenging. Using wild-type animals like mice, rats, chickens, or camelids often lowers upfront costs because no licenses or engineered animals are required, but the resulting antibodies must be humanized, adding significant risk, time, and cost. Humanized animals generate fully human variable regions at the outset, thus removing humanization requirements, but tend to be more expensive at initiation due to animal costs and licensing fees.

Display-based strategies have advantages with respect to throughput and lower initial costs, thus allowing for multiple antigen forms to be used in parallel or in series and increasing the likelihood of generating domain-specific binders or reducing cross-reactivity to closely related off-target sequences. In addition, display platforms can be critical tools for human sequences that are poorly immunogenic in mice and rats. Phage-based display platforms work with very large libraries but utilize a prokaryotic expression system. Yeast display campaigns leverage a eukaryotic expression system but start with medium sized libraries.

2.7.3.2 Immunization and Display-Selection Reagents

Creating and sourcing a robust set of target antigen reagents for animal immunizations or display library selection is critical, and most research teams do not invest enough time or money at this stage. It is not uncommon for discovery teams to commit significant capital and months of time to get to primary screen leads from a protein immunization campaign only to discover the original antigen was not folded properly. For each type of discovery reagent-like protein, DNA expression construct, or transgenic cell line, start with high quality and highly pure material that has been thoroughly tested to ensure the antigen will be presented in a biologically relevant format.

2.7.3.3 Screening Reagents and Assays

A well-characterized and diverse set of screening reagents and cell-based assays will be required at multiple stages of the discovery process. Early indications of success or failure come when polyclonal serum samples from immunized animals or enriched libraries from display panning are tested for binding to the target antigen and off-target controls. Later, with large panels of individual antibodies, primary and secondary screens for on- and off-target binding as well as functional characteristics will provide critical datasets for lead selection.

For protein and cell-based screening, it is valuable to collect and create reagents that represent multiple forms of the primary target antigen, often called orthogonal reagent sets, and relevant off-target controls. For example, if an animal is immunized with an Fc-tagged version of a protein target, consider checking serum titers and primary screen leads for binding to the original Fc-tagged immunogen, a

His-tagged version of the target, and an off-target protein with the same Fc tag to identify antigen-specific binders quickly and reliably.

2.7.3.4 Early Prioritization of Leads

Once you have individual antibodies that show the high-level binding properties of interest, focus can shift to prioritization of leads using assays for functionality. A detailed TPP will help guide which screening assays should be run in the earliest stages of prioritization. Historically, screening workflows tended to select for the highest affinity binders early, often at the cost of epitope, function, and sequence diversity. However, bioactivity may be more important than affinity for some projects, and the classic adage holds: You get what you screen for. Thus, it is often worth the investment to miniaturize the most biologically relevant functional assays for use in high-throughput screening.

Preclinical developability assays may be used as another prioritization factor, preferably in parallel with functional assays, to seek antibodies that are least likely to fail in later and more expensive stages of drug development. For therapeutic antibodies, desirable properties include high expression levels, high solubility, covalent integrity, conformational and colloidal stability, low polyspecificity, and low immunogenicity (Jain et al. 2017). No single assay result or series of assay results will be perfectly predictive of manufacturing and clinical success. However, prioritizing among a panel of lead antibodies with similar bioactivity profiles should favor those that raise the fewest warning flags across panels of biophysical assays.

2.7.4 Closing Thoughts

The antibody discovery and early screening process can be visualized as a funnel. The individual antibodies that come out of an immunization or display-based discovery campaign go into the top of the funnel and pass through multiple layers of filtering tests so that only the very best candidates come out at the end. Loading the top of the funnel with a large and diverse set of individual antibodies from a robust discovery campaign will ensure that high-quality leads remain even after attrition at multiple levels of filtering. In addition, including developability assays throughout the funnel layers will significantly increase the likelihood that the resulting lead antibodies are likely effective, manufacturable, and safe. These key concepts are illustrated in Fig. 2.6.

The Bottom Line

The best panels of therapeutic antibody candidates come from campaigns that fill the screening funnel with a large and diverse set of individual antibodies and then include filtering layers for both functional and developability characteristics.

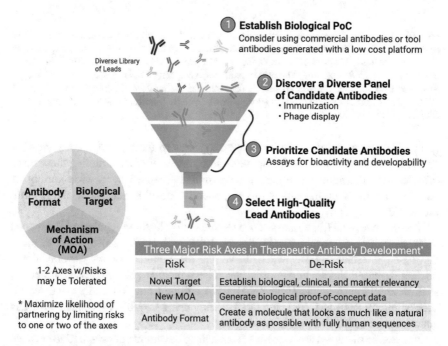

Fig. 2.6 Key concepts in the development of antibody therapeutics. Abbreviations: *MOA* Mechanism of Action, *PoC* Proof-of-Concept. (Created with BioRender.com)

2.8 siRNA, Gene Editing, and RNA Editing

Adriana A. Garcia, Rosa Bacchetta, and Jin Billy Li

Drug development has been dominated by small molecules and more recently antibodies, which elicit their desired effect by binding directly to a protein target. While these approaches have been successful for a variety of diseases, not every protein is druggable. Some lack defined ligand-binding pockets, three-dimensional structures,

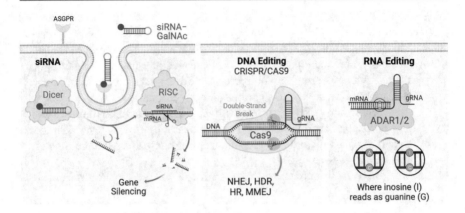

Fig. 2.7 The general mechanism for nucleotide-based therapeutic approaches. Abbreviations: *ADAR* Adenosine Deaminase Acting on RNA, *ASGPR* Asialoglycoprotein Receptor, *CRISPR* Clustered Regularly Interspaced Short Palindromic Repeats, *gRNA* Guide RNA, *HR* Homologous Recombination, *HDR* Homology Directed Repair, *mRNA* Messenger RNA, *MMEJ* Microhomology-Mediated End Joining, *GalNAc* N-Acetylgalactosamine, *NHEJ* Non-Homologous End Joining, *RISC* RNA-Induced Silencing Complex, *siRNA* Small Interfering RNA. (Created with BioRender.com)

or include a noncatalytic function that contributes to their pathogenicity, making antibody and small molecule therapeutics less than ideal when targeting these macromolecules. Furthermore, some diseases may be caused by a wide variety of genetic mutations in the target protein, making development of a single drug infeasible.

The discovery of RNA interference (RNAi) and clustered regularly interspaced short palindromic repeats (CRISPR), in 1998 and 1987, respectively, inspired cutting-edge technologies, which serve as the foundation for several novel nucleotide-based therapeutic approaches. These include gene silencing strategies, such as small interfering RNA (siRNA), RNA editing using RNA adenosine deaminase acting on RNA (ADAR), and DNA editing using CRISPR associated with Cas endonuclease (CRISPR/Cas) technologies.

Nucleotide therapies are attractive from a drug development perspective because they are not constrained by protein structure and can target any gene in the genome. The application of these technologies is not limited to therapeutic development as these methods are also used as tools in the drug development process. For example, siRNA can be used to identify and validate a given target in vitro and in vivo and DNA/RNA base editing can be used to establish proof-of-concept in animal models. This section will focus on the development of nucleotide-based therapies, with the general mechanism for each therapeutic approach illustrated in Fig. 2.7.

2.8.1   siRNA Therapeutics

The RNAi pathway is a biological process that regulates gene expression. This process is initiated when double-stranded RNA (dsRNA) binds to DICER, an endonuclease, which cleaves dsRNA into shorter, double-stranded fragments, known as siRNA and microRNA (miRNA). siRNA and miRNA form RNA-induced silencing complexes (RISCs) with argonaute proteins, triggering the degradation of complementary messenger RNAs (mRNAs) in a sequence specific manner.

Targeting the RNAi pathway is attractive from a therapeutic perspective as it may be used to silence genes involved in disease without editing the genome. siRNA is more promising than miRNA as a therapeutic as it is highly specific for its target sequence, whereas miRNA is more promiscuous. Consequently, siRNA-based therapies have been the developmental focus of many biotechnology companies. In 2002, Alnylam Pharmaceuticals obtained an exclusive license for a family of patents that covered RNAi-based therapies, granting them a tremendous advantage in siRNA-based therapies (Maraganore 2022). This has led Alnylam to develop three FDA-approved siRNA therapeutics: patisiran for transthyretin-mediated amyloidosis, givosiran for acute hepatic porphyria, and lumasiran for primary hyperoxaluria type 1 in 2018, 2019, and 2020, respectively. Alnylam has since formed alliances with other major biopharmaceutical companies and in December of 2021, Novartis developed the fourth FDA-approved siRNA therapy, inclisiran, to treat homozygous familial hypercholesterolemia. Alnylam recently received FDA approval for vutrisiran, a next generation siRNA treatment for transthyretin-mediated amyloidosis.

The past few years represent a turning point for siRNA therapies; however, many developmental obstacles had to be overcome and some hurdles remain. First, siRNA is unstable and short-lived in plasma, exhibiting a suboptimal pharmacokinetic profile that can reduce efficacy due to rapid elimination. Second, it is immunogenic, raising concerns regarding toxicity and loss of efficacy with subsequent dosing. Lastly, there are challenges in siRNA delivery, including targeting siRNA to specific cell and tissue types and cell/tissue penetration. Additionally, once it enters the cell, siRNA must escape from endosomes into the cytoplasm to perform its therapeutic function.

Chemical modification of the natural nucleotide sequence in siRNA has provided a successful strategy both for improving stability and specificity and reducing immunogenicity and toxicity. The most common modifications include replacing oxygen with sulfur in the linkage between nucleotides, known as phosphonothioate (PS) linkage, and modifying the 2′ position of the ribose sugar with 2′-OMe or 2′F. These modifications improve siRNA stability by increasing siRNA resistance to ribonucleases and improving their binding to plasma proteins, which in turn increases their time in circulation. Additionally, these structural modifications have been shown to improve affinity to the target mRNA, thereby reducing dosing requirements and associated toxicity of siRNA therapies.

Several siRNA modification patterns have been established by Alnylam and other siRNA-based companies and may be useful when developing siRNA therapies (Hu et al. 2020). These include standard template chemistry (STC), enhanced stabilization chemistry (ESC), and ESC+, with each generation having improved potency. Modifications rely on position-specific PS linkages and 2′ modifications, which increase stability and binding affinity of the molecule. These modifications provide the desired pharmacodynamic effect while limiting toxicity, as seen with successful development of givosiran, which relied on ESC. Additionally, such modifications have led to a second generation of patisiran (vutrisiran), which received FDA approval in June 2022.

Another common chemical modification is to conjugate siRNA to a ligand that directs siRNA to a specific tissue or cell type. For example, the asialoglycoprotein receptor (ASGPR) is highly expressed in hepatocytes, but not other cell types. Therefore, conjugating its endogenous ligand, N-acetylgalactosamine (GalNAc), to the siRNA results in a therapeutic that specifically targets the liver. This conjugation strategy was used in the development of givosiran, lumasiran, inclisiran, and vutrisiran, all having clinical indications related to liver disorders. Prior to this, patisiran was developed using nonconjugated ligand delivery. Here, siRNA was encapsulated by lipid nanoparticles. Once injected intravenously, apolipoprotein E (ApoE) associated with the siRNA-lipid nanoparticles and acted as a targeting ligand, binding to lipoprotein receptors on hepatocytes (Akinc et al. 2019). However, this method is inferior to GalNAc conjugation. By conjugating other receptor-specific ligands to siRNA molecules, the therapeutics may be targeted to other types of cells/tissues.

2.8.2   CRISPR and Therapeutic DNA Base Editing

DNA editing is the process of engineering a change in the nucleic acid sequence of DNA in a cell with high precision; this process can replace a single base pair or the entire coding sequence of a gene. Notably, genome editing constantly takes place in our cells through naturally occurring DNA breaks and repair, changing nucleic acid sequences to create functional diversity. For example, the immune system creates TCR (T cell receptor) and BCR (B cell receptor) diversity in the specificity of T and B lymphocytes, respectively, and readies them to fight infections against a virtually infinite number of antigens. In this example, the repair takes place via nonhomologous end joining (NHEJ). Another example is the diversity created by DNA recombination during meiosis, during which an exact copy of the DNA that was broken is copied and inserted at the precise site via homologous recombination, also called homology directed repair (HDR) (Carroll 2014).

We currently have tools available to artificially activate the DNA break and repair machinery at a desired site of the genome and to provide the desired sequence of the DNA to be inserted (Porteus and Baltimore 2003). For decades, gene editing technology involved the use of meganucleases, zinc finger nucleases (ZFNs), or transcription activator-like effector nucleases (TALENs) (Hirakawa et al. 2020). Progress in this field was greatly accelerated by the discovery and development of the CRISPR/Cas9 gene-editing system, where site specificity of endonuclease activity (by the Cas9 enzyme) is conferred by the complementary binding of guide RNA (gRNA) to genomic DNA. This combination is known as the Cas9-gRNA system (Porteus 2019).

The discovery of the CRISPR/Cas system made the editing process easy to design and suitable for use in different cell types (Jinek et al. 2012; Doudna and Charpentier 2014). During this process, the CRISPR/Cas editing complex finds and binds to a specific sequence in the DNA. The gRNA is complementary to the target DNA sequence and will bind only to the target sequence and no other regions of the genome. The Cas endonuclease follows the gRNA to the same location in the DNA sequence and makes a cut across both DNA strands. At this stage, the cell recognizes that the DNA is damaged and begins repair using endogenous DNA repair mechanisms. If a gene is to be inserted at the break site, this is usually delivered within a recombinant adeno-associated virus (rAAV) plasmid by transduction. Combining CRISPR/Cas9 with homologous recombination of a donor template delivered by rAAV6 enables highly efficient knock-in targeted integration, leading to therapeutic development in many disease applications that are reaching the clinic (Porteus 2019). Alternatively, methods of virus-free gene delivery of high amounts of DNA have also been tested to minimize inducing an antiviral type I interferon inflammatory reaction (Roth et al. 2018).

The CRISPR/Cas system can be employed for gene editing in many fields, such as in agriculture or microbiology. In medicine, this approach is a powerful tool that offers a cure for genetic diseases that are otherwise untreatable. Indeed, the CRISPR/Cas system can be exploited to insert new genes into precise sites of the genome, preserving the endogenous regulation of that gene; it can be used to delete genes or parts thereof or inactivate or mutate genes to acquire new functions. This gene manipulation for therapeutic purposes can be done on cells that are isolated from the patient from blood or tissues, edited in the laboratory, and then administered to the patient (ex vivo gene editing), or can be performed in vivo, by delivering the CRISPR/Cas system into the patient. In both cases, delivery is efficient, safe, and should not trigger an immediate immune response, which could inactivate the benefit of the editing. In ex vivo editing of hematopoietic stem cells, this risk is minimal because by the time the cell product is ready to be delivered to the patient, the CRISPR/Cas components are completely removed.

Bak and colleagues showed that combining CRISPR/Cas9 with rAAV6 delivery of a donor template enabled highly efficient knock-in targeted integration in the β-globin gene locus of hematopoietic stem and progenitor cells (HSPCs) (Dever et al. 2016). They further showed that blood progenitors from gene-corrected sickle cell patient HSPCs could produce healthy hemoglobin A, leading to a potential cure for sickle cell disease that is now undergoing clinical investigation. The donor template corrects the E6V mutation that causes sickle cell disease and includes six silent changes to prevent re-cutting. Other sequences at the locus remain unchanged. While the E6V mutation is characteristic of sickle cell disease, many monogenic blood disorders have patients with mutations spanning the gene. To encompass all patients regardless of mutation site, a donor template containing the entire complementary DNA (cDNA) of the gene can be inserted. The essential components of the donor are flanking arms of homology to the cut site, cDNA of the gene that has been codon optimized, a gene to mark edited cells, as well as other components to tune gene-specific expression such as introns, the woodchuck hepatitis virus post-transcriptional regulatory element (WPRE), and the 3' untranslated region. This approach to encompass all patient mutations has been explored in monogenic immune disorders, where patient mutations span the entire gene (Pavel-Dinu et al. 2019). We are currently exploring the gene correction by editing HSPC in patients with a rare genetic autoimmune disease. Because of the transcriptional activity of the mutated gene, which is highly regulated and expressed predominately in a very differentiated blood cell type called regulatory T cells and present in at least two different isoforms, we are testing vectors designed to preserve endogenous regulation and splicing. The system can achieve very high efficiency of editing (up to 70%) in HSPC. The question now being addressed is whether this is sufficient to provide a selective advantage of the edited cells vs. the unmodified ones while also minimizing the number of cells that are incorrectly edited.

Theoretical risks of treatment with ex vivo gene edited cells would be off-target double-strand breaks that could lead to cancer-causing mutations. Hendel et al. have successfully increased specificity and reduced off-target effects by decreasing the stability of gRNA using chemical modifications (Hendel et al. 2015). Off-target effects can be assessed for a given Cas/gRNA complex as well as in the final product by multiple in silico and deep sequencing methods. However, the cutoff for the number of off-target insertions/deletions (indels) for safety has not been established and will require well-monitored phase 1 clinical trials. Immunogenicity of gene editing components could also be a limitation for in vivo gene editing. Such a response could reduce the efficacy of additional treatments (Ferrari et al. 2021).

Clinical application of genome editing is already a reality for various medical indications and, to date, no major safety concerns have emerged. Several approaches are in the clinic for oncology, mainly with ex vivo generated chimeric antigen receptor (CAR) T cells (Qasim et al. 2017). For monogenic diseases in the blood and the immune system, HDR-mediated gene editing is a very promising treatment. Trials for sickle cell disease and thalassemia are likely to generate very important data for further improvement of current editing protocols. Combining newborn screening with gene editing of autologous hematopoietic stem cells could make several primary immunodeficiencies disappear. Moreover, autoimmunity and inflammation could be overcome by enforcing site-specific expression of regulatory genes and correcting specific cell subsets (Goodwin et al. 2020a). Likewise, alloreactivity could be prevented by NHEJ-mediated removal of alloreactive TCR. Gene edited cells could also be used to deliver a therapy to specific tissues in a more persistent and specific way than is possible with a traditional drug.

Overall, the editing technology provides hope for a new generation of medical treatments in an incredibly broad range of indications, providing that it is ethically performed to overcome incurable diseases. The big challenge for the future will be to make these therapies accessible to all patients in need (Gene therapies should be for all 2021).

2.8.3   Therapeutic RNA Base Editing

The goal of precision medicine is to furnish individualized therapies tailored to each patient's symptoms and genetic complement, delivering treatment with exquisite specificity and minimal side effects. The CRISPR/Cas system, including its variants such as base editors, was heralded for its efficiency and apparent specificity, but pervasive off-target effects have been revealed. Furthermore, adaptive immunity to Cas proteins derived from pathogenic bacteria may be widespread in patient populations (Charlesworth et al. 2019) with unknown consequences for efficacy and toxicity.

ADAR (adenosine deaminase acting on RNA) is a eukaryotic enzyme that edits RNA. Specifically, ADAR binds dsRNA and deaminates an adenosine base to inosine, which is then recognized as guanosine. There are two enzymatically active ADAR enzymes: ADAR1 and ADAR2. ADAR1 is highly expressed in most tissue types, perhaps because it provides a critical role in labeling cellular dsRNA as "self" to avoid recognition by the host innate immune sensor, MDA5 (melanoma differentiation-associated protein 5). ADAR2 expression is more tissue-specific, with higher levels in the brain and arteries.

ADAR provides an appealing system for site-directed RNA editing, also known as RNA base editing. A specific adenosine in an mRNA can be edited by delivering an antisense gRNA that forms a dsRNA structure with the target mRNA. This allows ADAR to bind and convert the adenosine to inosine, functionally providing an A-to-G base edit. Compared to DNA editing, RNA editing is transient and reversible. This may be viewed as disadvantageous because it requires repeated dosing to achieve an enduring treatment effect. However, it can also provide an advantage because it offers a better safety profile in case of adverse effects. Additionally, there are clinical scenarios where transient modulation of the activity of a target protein is desired.

The first generation of RNA base editing relies on the introduction of both the antisense gRNA and the deaminase domain (dd) of the ADAR enzyme. Assembly of the gRNA to the ADAR dd can be mediated by either covalent bond formation or noncovalent interactions. Approaches include: (1) forming a covalent bond between the ADAR dd and the gRNA (Stafforst and Schneider 2012); (2) fusing a SNAP-tag to the ADAR dd to form a covalent bond between the SNAP-tag and gRNA (the chemically modified gRNA can be conjugated to the SNAP-tag) (Vogel et al. 2018); (3) fusing an RNA-binding protein that has high affinity for a specific RNA sequence to the ADAR dd (the RNA-binding protein target sequence is coupled to a gRNA, resulting in a noncovalent interaction between the ADAR dd and the gRNA) (Montiel-Gonzalez et al. 2013); (4) fusing the RNA-targeting dCas13b enzyme (lacking nuclease activity) to the ADAR dd to achieve site-directed RNA editing similar to the aforementioned approaches (Cox et al. 2017). While highly efficient on-site editing is often achieved, off-target editing is extensive in these methods, perhaps due to the high expression level of the exogenous ADAR deaminase (Vogel et al. 2018).

The second generation of RNA base editing requires only delivery of the antisense gRNA to harness endogenous ADAR enzymes. As it does not deliver exogenous ADAR, this approach is extremely specific. The challenge largely lies in designing the gRNA so that upon forming dsRNA with the target RNA, it effectively recruits the endogenous ADAR enzyme for efficient site-specific RNA editing. In one design, the gRNA is heavily chemically modified, allowing the antisense sequence to be short (20–30 nt) with the optional recruitment domain of a hairpin dsRNA structure derived from a natural RNA editing substrate (~40 nt) (Merkle et al. 2019; Monian et al. 2022). The higher stability, specificity, and efficiency of the chemically modified gRNA oligonucleotide enables in vivo delivery using similar methods developed for siRNA and DNA antisense oligonucleotide (ASO) delivery (Monian et al. 2022). In another design, the much longer, genetically encoded gRNA (typically >100 nt) (Qu et al. 2019) allows for AAV delivery. One major challenge of this design is bystander off-target events. Multiple approaches have been developed to overcome this, including the CLUSTER design of recruitment sequences distributed over the target and the mismatches (Reautschnig et al. 2022), or bulges and mismatches introduced at undesired adenosines (Katrekar et al. 2022; Yi et al. 2022) to suppress nonspecific editing. Another challenge lies in the stability of long gRNA, which can be improved by circularization of the gRNA (Katrekar et al. 2022; Yi et al. 2022).

ADAR-mediated RNA base editing holds great potential for treating both rare and common diseases, although it is limited to making A-to-G changes. Compared to the two-component system of CRISPR/Cas-mediated DNA editing, RNA base editing is similar to siRNA and ASO, requiring delivery of a single, minimal component that harnesses endogenous cellular machinery. In addition to correcting pathogenic G-to-A mutations that are relatively common, RNA base editing can modulate gene activity much like small molecule and antibody therapies, but with unprecedented specificity. The emerging modality of RNA base editing will greatly benefit from improved oligonucleotide delivery methods developed in the siRNA and ASO fields, as well from advances in AAV-mediated delivery for a wider range of tissue accessibility.

The Bottom Line
Programmable gene editing holds immense therapeutic promises. While scientists continue to improve the technology, well-designed phase 1 studies are needed to better characterize safety and efficacy in patients, especially those with high medical need.

2.9   Cell and Gene Therapy

Maria Grazia Roncarolo, Alma-Martina Cepika, and Rosa Bacchetta

2.9.1   Introduction to Cell and Gene Therapy

Cells can be used as living drugs to deliver a desired product (*e.g.*, an enzyme), a gene (*e.g.*, gene addition in genetic diseases), or a desired function (*e.g.*, anti-tumor activity) to patients lacking those features. Thus, *cell therapy* can have enormous advantages over small molecules and biologics in diseases that require long-term, systemic delivery, including genetic, degenerative, or chronic inflammatory diseases, and cancer. Because of their ability to self-renew and travel to the target site, stem cells can be often administered as a one-shot, curative therapy. Cell therapies can also utilize mature cells such as T cells and NK (natural killer) cells.

The advantage of cells as living drugs has been recognized since 1959, when the first successful hematopoietic stem and progenitor cell transplant (HSPCT) was conducted between a healthy donor and her twin sister with leukemia. From 1959, the use of autologous (from the same individual) and allogeneic (from another donor) HSPCT continued to increase. Allogeneic HSPCT remains an important therapeutic option for many patients with genetic diseases or hematological malignancies. However, the immune barrier between human leukocyte antigen (HLA)-mismatched donor and recipient carries a high risk of graft-vs-host disease (GvHD). GvHD can be avoided by using an autologous HSPCT, where patients' own cells are manipulated ex vivo to restore a lacking feature or add an additional feature that could alter the course of the disease. This manipulation can consist of purification to obtain a desired cell subset (which can be subsequently expanded in vitro or recombined with other cells), in vitro differentiation to achieve a desired phenotype and/or function, or genetic modification. Genetic modification of ex vivo isolated cells, where genes are added using viral vectors or corrected using CRISPR/Cas9 or other gene editing platforms, is called ex vivo gene therapy. During in vivo gene therapy, patients receive an engineered, nonpathogenic virus that infects their cells and delivers the target gene. Collectively, we refer to these approaches as *cell and gene therapy*. The field of cell and gene therapy has been fueled by advances in molecular biology and medicine and has experienced a tremendous boom in the last two decades (Fig. 2.8).

The main categories of cell and gene therapies are as follows:

1. *Purified cell therapies*, which involve one or several purification steps of ex vivo isolated cells (e.g., allogeneic cord blood-derived CD34$^+$ HSPCs, or HSPCs depleted from TCRαβ$^+$ T cells and CD19$^+$ B cells before transplantation).
2. *In vitro modified cell therapies*, where cells are further manipulated in vitro after purification, but without genome engineering (e.g., mixture of cultured keratinocytes and fibroblasts in a collagen scaffold for the treatment of burns, or selectively expanded antigen-specific regulatory T cells for autoimmune diseases).

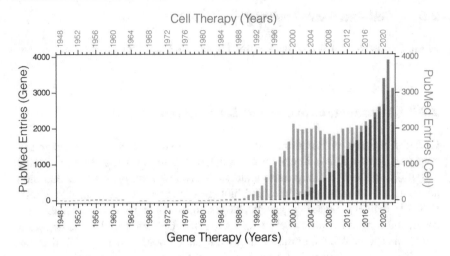

Fig. 2.8 Results of a PubMed search for the term *"cell therapy"* [Title/Abstract] (teal) or *"gene therapy"* [Title/Abstract] (grey). Current as of October 2022

Table 2.2 FDA- and/or EMA-approved ex vivo gene therapies for genetic diseases

Gene	Disease	Delivery	Name	Origin	Company
ADA	ADA-SCID	HSC transduced with a retroviral vector	Strimvelis	San Raffaele Institute, Milan, Italy	Orchard Therapeutics
HBB^{T87Q}	ß-thalassemia	HSPC transduced with a lentiviral vector	Zynteglo	BlueBird Bio, Cambridge, US	BlueBird Bio
ARSA	Metachromatic leukodystrophy	HSPC transduced with a lentiviral vector	Libmeldy	San Raffaele Institute, Milan, Italy	Orchard Therapeutics

Abbreviations: *ADA* Adenosine Deaminase, *ARSA* Arylsulfatase A, *HSPC* Hematopoietic Stem and Progenitor Cell, *HSC* Hematopoietic Stem Cell, *HBB* Hemoglobin Subunit Beta, *SCID* Severe Combined Immunodeficiency, *US* United States

3. *Gene therapies*, which have several subcategories:

   (a) *In vivo gene therapies with inactivated viral vectors* that contain a cassette with the target gene, which are administered directly to the patient either locally (e.g., adeno-associated virus (AAV) 2 carrying the retinoid isomero-hydrolase (RPE65) gene, injected into the subretinal space to treat congenital blindness) or systemically (e.g., AAV9 carrying the survival motor neuron 1 (SMN1) gene, delivered intravenously for treatment of spinal muscular atrophy).

   (b) *Ex vivo gene therapies with gene addition*, where the target gene is added to a cell of interest using a γ-retroviral vector or newer and safer lentiviral vector (for discussion on safety and examples see the text and Table 2.2).

   (c) *Ex vivo gene therapies with gene correction*, where the mutated target gene is corrected to its normal form by genome editing with CRISPR/Cas9 or other nucleases (e.g., hemoglobin subunit beta (HBB) gene correction for sickle cell disease caused by a mutation in hemoglobin B gene).

(d) *Chimeric antigen receptor (CAR) T cells*, which currently employ ex vivo gene addition to confer cancer-fighting properties to T cells (e.g., CD19 CAR T cells to treat CD19$^+$ B cell malignancies); it is listed here as a separate category because gene correction or the combination of gene addition and correction approaches for CAR T cells are also being explored.

2.9.2 Ex Vivo Gene Therapy

Major boosts to the field of ex vivo gene therapy using autologous cells have been brought by technological advances in whole genome sequencing, gene transfer technology, and clinical-grade in vitro cell manufacturing. However, the logistics of efficiently isolating, manipulating, and distributing personalized, autologous cell therapies in a timely manner present a major bottleneck. In addition, the clinical translation of cell and gene therapies frequently requires nontraditional and specific preclinical models, as opposed to typical toxicology, PK, and PD studies. Altogether, development of autologous cell and gene therapies carries a high amount of risk, which has not been readily accepted by major drug development companies.

This bottleneck has been successfully overcome in advanced academic medical centers, which can provide infrastructure for discovery, process development, manufacturing, regulatory affairs, and clinical trials to teams of basic, translational, and clinical scientists. As such, cell and gene therapy can represent an opportunity rather than an obstacle for academia. Indeed, successful ex vivo cell and gene therapies have been conducted with HSPCs to cure genetic diseases of the blood and immune system and even metabolic diseases associated with neurodegeneration. In addition, the field of cancer immunotherapy has recently experienced a rapid expansion of CAR T cells, which have been engineered ex vivo to target a tumor antigen and destroy tumor cells.

Several requirements must be met for the successful development and translation of cell and gene therapies from the bench to the bedside in academia:

1. *Scientific environment*, which consists of clinicians and scientists with insight into the mechanism, manifestations, and natural history of the disease. This expertise enables the design of appropriate preclinical studies, selection of the right genetic engineering approach and target cells, validated and informative in vitro and in vivo (murine or humanized mice) readouts for safety and efficacy, and selection of the appropriate study design with primary and secondary endpoints.

2. *Infrastructure* for sustaining the scalable and robust manufacturing process, which allows a successful tech transfer from the bench (small scale production) to the Current Good Manufacturing Practice (cGMP) laboratory, with defined and robust identity and efficacy release criteria. Infrastructure also includes a regulatory affairs team that can facilitate successful IND filings to FDA, as well

as a preclinical lab that can address questions or concerns raised by FDA and conduct correlative studies once the study has begun.

3. *Strategic support* from the university including funding to cover the costs of phase 1 clinical trials; clinical trial staff that are trained to help with patient recruitment, enrollment, and monitoring; manufacturing teams; regulatory support staff; and a technology transfer office that supports commercialization efforts with companies that are equipped for the financial and organizational challenges of phase 2 and 3 clinical trials and filing the BLA.

Because of the very dynamic state of the field, we will not present here a detailed overview of cell and gene therapies currently approved by FDA and European Medicines Agency (EMA). Instead, relevant links are provided in the Sect. 2.9 Resources below. In addition, a detailed overview of the development, applications, and challenges with HSPCT, CAR T cell therapies, gene addition, and gene correction approaches can be found in these reviews (Porteus 2019; High and Roncarolo 2019; Iglesias-Lopez et al. 2021; June et al. 2018; Majzner and Mackall 2018; Naldini 2019). Below, we will focus on our experience in ex vivo gene therapies for genetic diseases in the Center for Definitive and Curative Medicine (CDCM), founded in 2017 by Dr. Maria Grazia Roncarolo at Stanford University School of Medicine. CDCM has already yielded a spinout company, Graphite Bio, which has further advanced some of CDCM's pipeline.

2.9.3 Ex Vivo Gene Therapy for Genetic Diseases: The CDCM Experience

Genetic diseases are caused by deleterious germline alterations of the DNA sequence. These diseases are only rare when analyzed individually; collectively, 189–321 million people are currently living with a genetic disease (Nguengang Wakap et al. 2020). Patients with genetic diseases mostly depend on treatments that alleviate their symptoms, but do not address the cause of the disease. This often results in reduced life span or quality of life and can impose high healthcare costs together with significant socio-emotional and economic burden on the patient's immediate family. For example, treatment of patients that require frequent blood transfusions due to β-thalassemia, a disease of red blood cells caused by a mutation in the hemoglobin HBB gene, costs the US healthcare system between ~$150 and 275 million annually. Genetic diseases are thus an ideal target for ex vivo gene therapy, because adding or replacing the mutated gene with its wild-type version can provide a definitive cure.

In 2016, EMA granted the first marketing approval for an ex vivo gene therapy in a genetic disease. Strimvelis is a γ-retroviral vector carrying a cassette for the adenosine deaminase (ADA) gene for the treatment of ADA-SCID (severe combined immunodeficiency) (Aiuti et al. 2009). Strimvelis was not the first gene therapy in clinical trials; a similar γ-retroviral vector carrying the common γ-chain gene, IL2RG, was tested for X-linked SCID (Cavazzana-Calvo et al. 2000).

Although effective in treating SCID, this therapy was never approved because five out of twenty patients developed leukemia (Herzog 2010). As of early 2022, there are three ex vivo cell and gene therapies for rare diseases with marketing approval in the US and/or in Europe (Table 2.2).

CDCM's pipeline includes six ex vivo gene therapies in late preclinical or clinical stages (Table 2.3).

Table 2.3 Cell and gene therapies for genetic diseases in CDCM

Gene	Disease	Delivery	Clinical-Trials. gov ID	Transferred	Stage
Clinical					
COL7A1	Recessive dystrophic epidermolysis bullosa	Retroviral gene addition to keratinocytes	NCT04227106	Yes, to Abeona	Phase 3 clinical trial enrollment completed, awaiting marketing approval
S allele of *HBB* gene	Sickle cell disease	CRISPR/Cas9 gene correction of autologous HSPC	NCT04227106	Yes, to Graphite Bio	Phase 1/2 clinical trial halted
FOXP3	IPEX syndrome	Lentiviral gene addition to CD4+ T cells	NCT05241444	No	Phase 1 clinical trial
Preclinical					
MPS1	Type I mucopolysaccharidosis	CRISPR/Cas9 gene correction of autologous HSPC	–	–	Pre-IND
IL2RG	SCID-X1	CRISPR/Cas9 gene correction of autologous HSPC	–	–	Pre-IND
CFTR	Cystic fibrosis	CRISPR/Cas9 gene correction of autologous airway stem cells	–	–	Late preclinical

Abbreviations: *CRISPR* Clustered Regularly Interspaced Short Palindromic Repeats, *COL7A1* Collagen Type VII Alpha 1 Chain, *Cas9* CRISPR Associated Protein 9, *CFTR* Cystic Fibrosis Transmembrane Conductance Regulator, *FOXP3* Forkhead Box P3, *HSPC* Hematopoietic Stem and Progenitor Cell, *HBB* Hemoglobin Subunit Beta, *IPEX* Immune Dysregulation, Polyendocrinopathy, Enteropathy, X-Linked, *IL2RG* Interleukin 2 Receptor Subunit Gamma, *IND* Investigational New Drug Application, *MPS1* Mucopolysaccharidosis Type 1, *SCID-X1* X-Linked Severe Combined Immunodeficiency

2.9.4   The Story of IPEX

Notably, several different platforms for gene therapy of the same target can be explored at the same time, providing there is strong scientific rationale. This is well illustrated in the development of ex vivo gene therapy for immune dysregulation, polyendocrinopathy, enteropathy, X-linked (IPEX) disease, a severe monogenic autoimmune disease. This development is spearheaded by Dr. Rosa Bacchetta, a pediatric immunologist, former SPARK grant recipient, and Associate Lead for CDCM's "Engineering the Immune System" platform.

Forkhead Box P3 (FOXP3), the gene mutated in IPEX, is a transcription factor critical for differentiation of regulatory T cells (Tregs) that prevent autoimmunity. IPEX is naturally modeled in mice, where a mutated FOXP3 gene results in the *scurfy* phenotype and severe autoimmunity. Murine FOXP3 is specifically expressed in regulatory T cells, whereas human FOXP3 also has a secondary role in other immune T cells, as the result of a more complex endogenous regulation of expression in the two cell types. These species-dependent and cell-subset-specific differences in biology pose a challenge in translating the therapeutic strategy tested in *scurfy* mice to patients with IPEX and have revealed humanized mice as a better preclinical model to test possible gene therapy approaches for this disease (Goodwin et al. 2020a; Sato et al. 2020). Thus far, a lentiviral vector delivery strategy of the FOXP3 gene and its promoter in HSPCs of *scurfy* mice has been successful in restoring Treg lineage-specific FOXP3 expression (Masiuk et al. 2019). On the other hand, human HSPCs transduced with lentiviral vector-mediated enforced expression of FOXP3 displayed a proliferative defect in all T cells (Santoni de Sio et al. 2017). However, since the main symptom of patients with IPEX is autoimmunity, for which regulatory T cells are key players, an alternative approach was to use lentiviral vector-delivery with a constitutive promoter to enforce FOXP3 expression in CD4[+] T cells. This approach was also successful in preclinical humanized mice models (Sato et al. 2020; Passerini et al. 2013), leading to the recently commenced phase 1 clinical trial with an engineered regulatory T cell product called CD4[LVFOXP3]. Dr. Bacchetta and CDCM also continue to develop a CRISPR/Cas9 gene correction therapy for IPEX, which will replace the mutated FOXP3 in human HSPCs with its wild-type version, preserving endogenous regulation and allowing differentially regulated expression in Treg cells and other CD4[+] T cells (Borna et al. 2022). This approach has shown promising preclinical results (Goodwin et al. 2020b) and is now in late-stage preclinical development.

The gene therapy for IPEX story also illustrates the importance of having infrastructure in place to foster collaboration among clinicians, basic, and translational scientists. Recruitment and diagnosis of patients with IPEX is facilitated by Stanford's Center for Genetic Immune Diseases (CGID), which consists of physician scientists specializing in pediatric hematology, immune deficiencies, and stem cell transplantation, along with a dedicated research laboratory. CGID aids the diagnosis and recruitment of patients with suspected genetic immune diseases such as IPEX. In addition, development of the CRISPR/Cas9 gene correction approach for IPEX greatly benefited from collaboration with the laboratory of Dr. Matthew Porteus, a leader in the field of CRISPR/Cas9 gene correction strategies for various diseases. Finally, the technology transfer, process development, cGMP manufacturing, IND filing, and fundraising to cover the cost of the phase 1 clinical trial were heavily dependent on a large team of experts from Stanford research groups, the CGID clinic, Division of Stem Cell Transplantation and Regenerative Medicine, Clinical Trials Office, Laboratory for Cell and Gene Medicine (Stanford's cGMP facility), and CDCM's research administration office. Substantial funding from the California Institute for Regenerative Medicine (CIRM) proved instrumental in supporting preclinical development and phase 1 trials of many CDCM projects, including that of IPEX.

Finally, the IPEX example also highlights the therapeutic power of regulatory T cells, which are an integral part of our immune system and prevent autoimmune and autoinflammatory diseases. As such, regulatory T cells are implicated in many human diseases and therapeutically have been leveraged in transplantation and autoimmunity (Ferreira et al. 2019; Sayitoglu et al. 2021). The prime indications for ex vivo gene therapies that correct deficiencies in regulatory T cells are Tregopathies (Cepika et al. 2018), autoimmune diseases caused by mutations in genes critical for regulatory T cell survival and function. But we believe that lessons learned in regulatory T cell therapy for IPEX and other monogenic autoimmune diseases may lead to broad applications of these therapies in polygenic autoimmunity, inflammation, and allergy.

Altogether, we hope that the examples provided herein will encourage more academics to translate their knowledge and skills into new cell and gene therapies and provide a framework of necessary ingredients that an academic medical center needs for successful translation of cell and gene therapies from the bench to the bedside. With the rapid advancement of gene engineering technology, we are on the cusp of a prolific era of personalized medicines.

Resources

- *US Food and Drug Administration (FDA)* website listing currently approved cell and gene therapy products: https://www.fda.gov/vaccines-blood-biologics/cellular-gene-therapy-products/approved-cellular-and-gene-therapy-products.
- *European Medicines Agency (EMA)* website for advanced therapy medicinal products (ATMPs; term used for cell and gene therapies in Europe): https://www.ema.europa.eu/en/human-regulatory/overview/advanced-therapy-medicinal-products-overview.
- *Alliance for Regenerative Medicine*, an international advocacy organization supporting advanced therapies and regenerative medicines that issues a highly informative annual report: https://alliancerm.org/.
- *California Institute for Regenerative Medicine (CIRM)*, which strives to accelerate new therapies to fill unmet medical needs, including provision of funding for basic, preclinical, and clinical research studies: https://www.cirm.ca.gov/.
- *American Society for Cell and Gene Therapy*, which has an annual meeting and educational material for its members: https://asgct.org/.
- *Center for Definitive and Curative Medicine (CDCM)*, Stanford University's Center for Definitive and Curative Medicine: https://med.stanford.edu/cdcm.

2.10 Vaccine Development

Harry Greenberg

Few, if any, biomedical interventions have been as successful at preventing human morbidity and mortality as vaccines. The eradication of smallpox, the near eradication of paralytic polio, and the substantial reduction of the global burden of hepatocellular and cervical cancer are just a few of the many benefits rendered by vaccines in the last 50 years. Vaccines are also one of the most egalitarian of all health interventions, since their benefits generally are well suited for delivery to both wealthy and poor countries alike. Therefore, vaccines have the ability to rapidly and efficiently alter the face of global health and well-being.

Vaccines are molecular moieties (antigens or RNAs encoding antigens) that can be administered via several routes, such as parenterally (*e.g.*, intramuscular, subcutaneous, intradermal) or via a mucosal surface (*e.g.*, orally or intranasally). In general, they are administered on only one or a few occasions because they are designed to elicit a long-lasting immune response in the host. However, in some cases, vaccines are administered more regularly when the target antigen changes at regular intervals, such as influenza. They can be formulated from simple proteins or peptides, polysaccharides, protein-encoding nucleic acids, or complex mixtures of these constituents. In addition, vaccines can be created using various target infectious agents that are attenuated in some fashion and whose replication is substantially restricted. These infectious agents can, on occasion, also be used to carry and express exogenous proteins.

The most successful vaccines have generally been live attenuated infectious agents, inactivated infectious agents, or complex components of infectious agents or polysaccharides conjugated to protein carriers. As well demonstrated for the SARS CoV-2 vaccine, the delivery of lipid droplet-encased RNA encoding specific target antigens has been added to our strategies to produce highly effective vaccines. Vaccines are employed to induce a host immune response that is either protective or therapeutic. Thus far, vaccines have been substantially more effective as preventatives to avoid contracting the disease than as therapeutics to treat a disease or infection. The general or even specific applicability of the "therapeutic vaccination" concept remains to be better defined in humans.

Vaccination has been most successfully employed to prevent a wide variety of infectious diseases caused by many different viruses and bacteria. Vaccination against parasitic diseases has been much less successful thus far. "Vaccination" or immunization with specific allergens, especially food allergens, has also been used with some recent success for allergy treatment. In addition, a variety of experimental vaccines to treat substance addiction, for birth control, and to treat autoimmune diseases have been studied but have not yet been generally successful. This section will therefore focus specifically on preventative vaccines against infectious diseases.

2.10.1 Vaccine Efficacy

The past 70 years have witnessed the development of many highly successful new vaccines. The remaining important infectious disease targets, such as HIV, tuberculosis, malaria, cytomegalovirus (CMV), and several common sexually transmitted diseases (STDs), remain difficult to prevent. Vaccine development has the highest likelihood of success when the natural infection induces a strong and enduring immunity to subsequent infection or illness. This was the case for smallpox, measles, and hepatitis A and B. In other cases where reinfection can occur (usually at a mucosal surface) but the secondary infection is generally as severe as the initial one, vaccination approaches have also been successful. This is the case, for example, with rotavirus and influenza vaccines.

When one or a few natural infections do not lead to development of significant immunity—such as HIV, hepatitis C virus (HCV), gonorrhea, and rhinovirus—it is likely that the pathway to an effective vaccine will be far more difficult. In these cases, identification of novel immunization strategies is likely required to develop a successful vaccine.

Two key elements to facilitate successful vaccine development are the availability of a simple predictive assay to quantify a protective vaccine immune response and a relevant animal model in which to test the efficacy of various immunization strategies. Animal models that replicate actual wild-type infections of the microbial pathogen in the human host are most likely to be relevant. The duration, specificity, and strength of the host response, as measured by a validated functional assay, are frequently key determinants of the vaccine's efficacy.

2.10.2 How Vaccines Generally Work

Vaccines are designed to induce the host to mount an immune response that prevents, reduces, or eliminates infection by the targeted pathogen. The induction of host immunity involves a variety of factors, including many aspects of the innate and acquired immune systems, the site of immune induction, the nature of the antigen and frequently an accompanying adjuvant, and the quantity and duration of antigen exposure. Each of these aspects needs to be carefully considered to maximize the chances of eliciting an acquired antigen-specific immune response that has functional therapeutic activity.

Whereas both T and B cell responses are commonly induced by vaccination, most successful vaccines "work" at the effector level primarily on the basis of B cell and induced antibody responses. Of course, T cell responses are critical to most effective B cell responses, and T cells frequently also play an effector role in generating immunity, especially related to infection resolution. The general mechanism for how vaccines work is illustrated in Fig. 2.9. Many methods have been and are being examined to enhance the immune response to vaccines, including using an adjuvant to boost the innate immune response, using protein carriers to induce immune memory to polysaccharide antigens, using multivalent, particulate protein

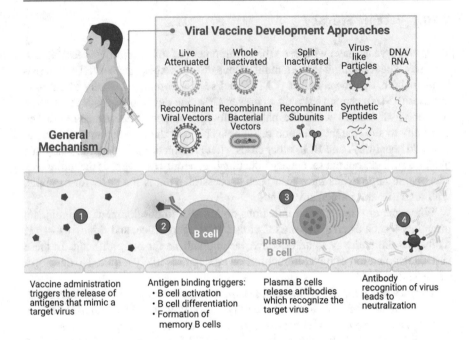

Fig. 2.9 General mechanism and development approaches for viral-based vaccines. Generally, viral vaccines work by delivering or causing the production of antigens that mimic a target virus. This causes naive B cells to differentiate into both plasma B cells that produce antibodies against the target antigen and memory B cells that can be activated upon re-exposure to the virus. T cells play a role in B cell activation (not shown) and in some cases have a direct antiviral effect that restricts replication. (Created with BioRender.com)

immunogens to enhance immunogenicity, and using replicating or single cycle non-replication competent vaccines to produce more antigens with greater diversity at the site of infection (viral-based vaccine approaches are summarized in Fig. 2.9). As mentioned above, when natural infection induces protective immunity, it has generally been relatively straightforward to design a vaccine that mimics the effective component(s) of that infection. When natural infection is not an effective inducer of protective immunity, vaccine development has been much more difficult.

2.10.3 Some New Technologies in Vaccine Development

This section cannot begin to cover all the new technologies that are currently being explored to develop novel or improved vaccines. Therefore, a few examples are provided to invite you to examine the field more extensively. Many pathogens avoid host immunity by altering or expanding their antigenic diversity. Examples include diverse organisms such as influenza, HIV, pneumococcus, and SARS-CoV-2. Recent advances in immunology have demonstrated the existence of "common" or "shared" antigens on several pathogens, such as identifying the influenza hemagglutinin antigen (HA) stalk as a target of a protective antibody. Such targets could provide an "Achilles' Heel" to which the host can target its immune response and thereby circumvent the problem of pathogen antigenic diversity. Many investigators are currently working to design new vaccines directed at shared common antigens for these infectious agents.

As an alternate approach, directed regulation of the innate immune response holds promise to greatly enhance the level and duration of acquired immunity following vaccination. Many investigators are now exploring the safety and efficacy of new adjuvants that directly target specific signaling molecules, thereby enhancing the innate immune response.

Finally, immunization using selected nucleic acid sequences (DNA, or more recently, RNA) encoding antigenic proteins holds great promise, as recently unequivocally demonstrated with the highly successful launch of two COVID-19 vaccines, both proven to be highly safe and effective within about 1 year of inception. Continued innovation in the area of RNA vaccination is likely to greatly facilitate and accelerate vaccine development in the coming decade (Hogan and Pardi 2022).

> **Box 2.11: What Surprised an Academic?**
> Until the recent COVID-19 pandemic, basic biomedical academics, physician-scientists and drug regulators generally anticipated that the development of a new vaccine would take at least a decade or more. To the pleasant surprise of all, the use of a novel RNA vaccine technology, along with the rapid use of various crystallographic techniques, enabled the development and implementation of highly effective vaccines for COVID-19 in under a year! —*HG*

2.10.4 Special Considerations Concerning Safety and Cost

A variety of factors distinguish vaccine development from virtually all other areas of therapeutic development. Of course, like all other medical interventions, vaccines must be shown to be efficacious. However, unlike most other interventions, vaccines are generally given to healthy individuals, often very young children, with the intent of preventing a possible illness in the future rather than treating a current problem. Because of this, the level of tolerance for risk associated with vaccination is dependent on the level of perceived danger from the infection being prevented. For example, when polio epidemics were common, the public clamored for a preventative intervention. However, since polio has disappeared from the Western hemisphere, even one case of immunization-induced polio per million vaccinations represents an unacceptable risk in the US and Europe. The very vigorous debate and public concern about the appropriate use of vaccination as a societal intervention to prevent COVID-19 is a striking example of the complexity of vaccine policy in many countries.

This common and pervasive concern with vaccine risk is often intensified because vaccines are most frequently given to young healthy children, who can be considered most vulnerable to untoward risk. In addition, the benefits of vaccination are most easily measured at the societal rather than the individual level, because the odds that any given individual will be infected are often quite low. This dichotomy further complicates acceptance of vaccines by the public. Because of these factors, vaccine development often requires investment in very large and extensive safety testing before registration, as well as substantial postlicensing follow-up that is both expensive and complex.

Because vaccines are most often given to healthy individuals, are generally given only a few times during the life of an individual, and frequently have a prolonged regulatory pathway due to safety concerns as discussed above, they have often been perceived as providing a poor return on investment by drug developers. This is, of course, a shame, given their immense societal impact over the years.

Finally, many important remaining challenges in vaccine development involve diseases (HIV, tuberculosis, malaria) that generally afflict poor, disadvantaged, and less-developed regions of the world. This fact has likely inhibited the rate of progress for these much-needed interventions. Despite these issues, recent advances in immunology, material sciences, systems biology, and nucleic acid-based immunization provide exciting opportunities for vaccine innovators of the future. During the coming decade, we are likely to see vaccination for several of these challenging diseases reduced to practice.

Box 2.12: What Surprised an Academic?
Unless the targeted disease is quite prevalent, a large number of patients must be included in vaccine trials to demonstrate efficacy—even for a highly effective vaccine. This can greatly add to development costs and trial duration. —*HG*

The Bottom Line
Although only one or a few doses of a vaccine are administered, vaccines are generally administered to healthy people, most often children, who are at low risk of acquiring the disease. As a result, the safety hurdle is very high, often adding to the time and cost of vaccine development.

2.11 Diagnostic Biomarkers

Steven N. Goodman and Michael A. Kohn

Biomarkers are important in clinical medicine, medical research, and drug development. An FDA/NIH Working Group paper (FDA-NIH Biomarker Working Group 2021) defines a biomarker as:

> A characteristic that is measured as an indicator of normal biological processes, pathogenic processes, or biological responses to an exposure or intervention, including therapeutic interventions.

Biomarkers are distinct from direct clinical outcome assessments such as pain, functional impairment, or death. Biomarkers can have many uses, which the Working Group paper categorized as:

1. Susceptibility/risk biomarker
2. Diagnostic biomarker
3. Monitoring biomarker
4. Prognostic biomarker
5. Predictive biomarker

A unifying framework for all of these biomarker types is through the lens of risk prediction and decision making. All biomarkers, to be useful, must modify the probability estimate for some outcome, whether that be disease risk (for susceptibility), disease presence (for diagnosis), disease course (for monitoring), disease outcome (for prognosis), or treatment effectiveness (for predictive biomarkers). A companion diagnostic is a type of predictive biomarker. These modified probabilities must cross a threshold that would justify a different clinical action by a caregiver or behavioral change by a patient; if biomarkers do not alter actions, they cannot improve outcomes. This is known as the threshold theory of decision-making.

We will focus here on diagnostic biomarkers used to detect the presence of a disease or condition of interest, but many of the principles apply to the other uses through the risk prediction paradigm. The main use of such biomarkers is to guide decisions about treatment or further testing with the aim of improving patient outcomes. The decision guidance means that the biomarker's assessment must be decision-specific, and the goal of outcome improvement means that measuring test accuracy, that is, sensitivity and specificity, is only the first step in a longer process.

Although a disease is not necessarily binary, we divide individuals into those with it (D+) and without it (D−). D+/D− means "disease positive/negative" where "disease" refers to a pathogenic process requiring a different course of treatment or monitoring. D+/D− could also mean "decision positive/negative," a better framing and one more applicable to the wide range of biomarker uses.

The steps in developing a diagnostic biomarker are to establish the following:

1. Reliability: Analytic reproducibility across different conditions
2. Phase 1 Clinical Validity: Differentiation between D+ and D− sampled separately
3. Phase 2 Clinical Validity: Differentiation between D+ and D− in the target clinical population
4. Clinical Utility: Improved decisions and clinical outcomes in the target clinical population.

2.11.1 Reliability

A biomarker measurement should give the same or similar results when made repeatedly on the same individual within a time too short for real biological variation to take place. This characteristic is called "reliability." Reliability encompasses not only technical variability but also real-world variations in sample acquisition, preparation, and transport. It is important to separate analytic from diagnostic accuracy (Saah and Hoover 1997). Statistical approaches to assessing reliability include coefficients of variation, Bland-Altman plots, and Kappa statistics.

2.11.2 Clinical Validity

2.11.2.1 Statistical Assessment of Test Accuracy

For a diagnostic biomarker to be useful, it must give different results in diseased (D+) and nondiseased (D−) individuals. This is assessed in a diagnostic test accuracy (DTA) study. A DTA study consists of an index test (the biomarker), a reference standard to determine D+/D−, a study population, and a sampling frame. We will assume, for the moment, that the biomarker is used along with a cutoff to separate a positive index test (T+) from a negative index test (T−). When both the index test and the disease are dichotomous, sensitivity, $Pr(T+|D+)$ and specificity, $Pr(T−|D−)$ are appropriate metrics of test accuracy. (The "|" symbol represents conditional probability and is read "given.")

Box 2.13: Measures of Diagnostic Test Accuracy
Disease based on the reference standard:

   D+ = presence of the target condition/disease
   D− = absence of the target condition/disease

Binary index test (biomarker with cutoff):

   T+ = index test positive
   T− = index test negative

Conditional probabilities
Known disease status

   Sensitivity = $Pr(T+|D+)$ = Probability of a **positive** index test given **disease**
   Specificity = $Pr(T−|D−)$ = Probability of a **negative** index test given **non-disease**

Known test result

   Positive Predictive Value = $Pr(D+|T+)$ = $Pr(\textbf{disease})$ given a **positive index test**
   Negative Predictive Value = $Pr(D−|T−)$ = $Pr(\textbf{non-disease})$ given a **negative index test**

While these properties are often illustrated with the classic 2×2 table, such tables offer little insight into how these indices are related. Bayes's theorem offers a more unified treatment through the prediction paradigm.

Box 2.14: Bayes's Theorem

$$\text{Posttest odds}(D+) = \text{Pretest odds}(D+) \times \underbrace{\frac{\Pr(\text{Test result} \mid D+)}{\Pr(\text{Test result} \mid D-)}}_{\text{Likelihood ratio}}$$

$$\text{Odds}(x) = \frac{\Pr(x)}{1 - \Pr(x)} \qquad \Pr(x) = \frac{\text{Odds}(x)}{1 + \text{Odds}(x)}$$

Note that in the above equation, "Test result" can be a continuous biomarker value or dichotomous. If the test is dichotomous, the likelihood ratio becomes Sens/(1−Spec) when T+ and (1−Sens)/Spec when T−. This shows clearly that the role of the test is to modify the probability of an outcome, that the pretest odds play a critical role, and the effect of pretest odds is separable from that of the test accuracy, which exerts its effect through the likelihood ratio.

A numerical example highlights this: assume a test with 90% sensitivity and 90% specificity. We apply it to a population with a 10% probability of disease. What is the probability of that condition after a positive test? Pretest odds = 0.10/(1 − 0.10) = **1/9**. Likelihood ratio = 90/(100 − 90) = **9**. Applying Bayes's Theorem, Post-test odds = 1/9 × 9 = **1**. Post-test probability = 1/(1 + 1) = **0.50**.

Thus, the post-test probability of disease can differ dramatically from that naively suggested by the diagnostic test accuracy (90%) and depends critically on the prevalence of that condition in the target population. The threshold probability for the decision depends on the clinical specifics.

Box 2.15: What Surprised an Academic?

Researchers who are inexperienced in developing diagnostic tests often naively believe that a test with high sensitivity and high specificity will be of clinical value. A colleague who wanted to develop a saliva test as screening tool for asymptomatic oral cancer approached us for assistance. Based upon sampling of saliva samples from several hundred patients with oral cancer and healthy controls, the calculated sensitivity and specificity were both 95%. Given a historical annual oral cancer rate of 10 new cases per 100,000 individuals in the population, the predictive value of a positive result would be (10 × 0.95)/[(99,990 × 0.05) + (10 × 0.95)]. Fewer than 2/1000 positive tests would have oral cancer and the test would not have clinical utility as a screening test. On the other hand, perhaps the test could be developed to monitor for disease recurrence following treatment in oral cancer patients whose test was positive at the time of diagnosis. The utility of a diagnostic test is dependent upon the patient population undergoing the test. —KVG

2.11.2.2   Sampling Frame

DTA studies can use case-control, clinical population-based, or test result-based sampling.

- *Case-control*: In case-control sampling, D+ and D− individuals are identified and sampled separately, such as taking D+ individuals from a disease registry and D− individuals from a primary care clinic. The investigators choose how large the D+ and D− groups are. DTA studies using case-control sampling should not naively calculate and report positive predictive value and negative predictive value, because these properties are determined by disease prevalence, which is determined by the investigator. If the separate samples are representative of the D+ and D− groups in the clinical population, sensitivity and specificity will be valid (but see below about spectrum bias). This type of sampling is used in phase 1 Clinical Validity studies.
- *Clinical population-based*: A clinical population-based sample is defined by the clinical characteristics of the population for which the test is intended; the "target population." Sensitivity and specificity estimates will be valid. The proportion of the sample with the disease, Pr(D+), is also a valid estimate of the prevalence of disease in the clinically relevant population, so positive predictive value and negative predictive value will be valid. This type of sampling is used in phase 2 Clinical Validity studies.
- *Test result-based*: In test result-based sampling, T+ and T− individuals are sampled separately and then both subjected to the reference test. This is the converse of case-control sampling. An example is taking all patients who had a positive cardiac stress test (T+), a random sample of those who tested negative (T−) and subjecting them all to cardiac angiography, the reference standard for coronary artery disease. If the separate samples are representative of the T+ and T− groups in the clinical population, the positive and negative predictive values will be valid. These predictive values can be used together with the sampling fractions to estimate sensitivity and specificity. This type of sampling is used in a phase 2 Clinical Validity study when it is impractical or unacceptable to apply the reference standard to all tested individuals, particularly those whose index test is negative.

2.11.2.3   Bias in Studies of Clinical Validity

Phase 1 Clinical Validity studies of biomarkers that use case-control sampling often overestimate both sensitivity and specificity. They overestimate sensitivity because the D+ cases often are more severe and homogeneous than D+ individuals in the clinical population. They overestimate specificity because the D− individuals are healthier than individuals who would be tested, who typically have signs and symptoms suggestive of the condition being tested for. This effect is called *spectrum bias*.

Phase 2 Clinical Validity studies use samples from the clinical population. When basing the study on clinical records, it is common and tempting to sample patients who, in the course of standard clinical workups, received both the index and the reference test. For example, a study might sample from all those who received cardiac angiography after a preceding stress test, rather than from all those who underwent the stress test. However, sampling only those who received the reference standard produces *verification bias*, which can dramatically overestimate sensitivity and underestimate specificity if not properly adjusted for, and sometimes adjustment is not possible (Kohn 2022a).

Phase 2 Clinical Validity studies that use samples from the clinical population can be subject to many other biases (Kohn et al. 2013). However, if the individuals are sampled from a realistic target population and the reference standard is valid and applied *independently of the index test result*, metrics of test accuracy should be valid.

2.11.2.4   Continuous Tests

We have thus far assumed that the biomarker is used with a cutoff that makes it a dichotomous test. But biomarker measurements are often continuous, or at least ordinal (*e.g.*, normal, suspicious, abnormal), and choosing a single cutoff between positive and negative can reduce the value of the biomarker. The ability of a continuous biomarker to discriminate between D+ and D− individuals can be summarized using a receiver operating characteristic (ROC) curve and the area under it (AUROC). Sometimes the continuous range of test results can be usefully divided into several intervals, just as BMI, systolic blood pressure, and blood glucose are divided into intervals (Kohn 2022b).

The optimal cutoff is the one that minimizes patient harm from test errors, a calculation that involves the absolute number of false positives and false negatives in the target population times the harms of each error. The ROC curve, which is often used to choose a cutoff, has no information about either absolute numbers or harms, so should not be used for that purpose. Instead, cutoffs should be chosen with equations that calculate net benefit (Newman and Kohn 2019).

2.11.3 Clinical Utility

Even if a diagnostic biomarker accurately distinguishes between D+ and D− individuals, it will only improve outcomes if it leads to more effective or less harmful interventions than would have occurred otherwise, that is, it has clinical utility. One way to determine this is to compare outcomes in groups randomized to having or not having the biomarker measured, which are unfortunately rarely done (Lord et al. 2006).

Instead, clinical utility is often determined with observational data. This requires specifying exactly where in the clinical decision pathway the biomarker is measured, what decision is to be affected, what the decision would have been with or without the measurement, and what the clinical benefit of that decision would be. This is where many biomarkers fail; even if their accuracy is good when considered alone, when combined with other information they often do not provide sufficient incremental value to justify their cost or burden.

Observational studies of biomarkers' clinical utility can include before-after studies, studies that use regional or site practice variation to create comparison cohorts, or within-site studies with statistical control for covariates. These often must be combined with external data on intervention efficacy. Such assessments are also done using existing datasets augmented by formal modeling. While FDA does not usually require demonstration of clinical utility of biomarker-based tests, payers often do.

2.11.3.1 Reporting Guidelines and Quality Assessment

Comprehensive reporting guidelines for DTA and prognostic test studies, Standards for Reporting Diagnostic Accuracy Studies (STARD) and Transparent Reporting of a multivariable prediction model for Individual Prognosis Or Diagnosis (TRIPOD), respectively, can be found on the EQUATOR website (www.equator-network.org). QUADAS-C is a risk of bias tool for DTA studies (Yang et al. 2021).

2.11.4 Conclusion

The purpose of measuring biomarkers in a clinical setting is to improve patient outcomes. Biomarkers aim to do this through modifying the probability estimates for various outcomes, either alone or combined with other clinical and test information. This in turn must affect clinical or patient actions in a manner that provides net clinical benefit for a specified target population. The assessment of that benefit, with its attendant uncertainty, provides the justification for clinical use and for reimbursement, and can facilitate FDA approval.

Box 2.16: What Surprised an Academic?
FDA regulates diagnostic tests as devices, with different (and lower) evidential standards than for drugs, albeit with requirements that vary for devices with different risks levels. Also, a test where a patient sample is sent to a central laboratory for measurement, aka a "Laboratory Developed Test", is subject to yet less FDA oversight. —*SNG* and *MAK*

Key Terms and Abbreviations

Key Terms

Adenosine Deaminase Acting on RNA (ADAR): A eukaryotic enzyme that edits RNA and can be used for therapeutic RNA editing.
Adjuvant: Compound that increases the host immune response to an antigen.
Antigen: Entity that activates an immune response in a host.
Attribution: Assigning causative value to an input data point for an observed output measurement.
Big Data: Datasets that are not easily understood through human inspection; they can scale from hundreds to millions of entries.

Biological Outputs: Measurements of a biological system (*e.g.*, assays or standard cellular responses).

Chemical Hit: Small molecule that affected the target or phenotype.

Chemical Moiety: A functional group or portion of a molecule.

Clustered Regularly Interspaced Short Palindromic Repeats (CRISPR): A family of sequences evolved in bacteria as a primitive immune system.

CRISPR Associated Protein 9 (Cas9): A DNA endonuclease enzyme that can cut DNA in a precise location complimentary to its guide RNA.

CRISPR Associated with Cas Endonuclease (CRISPR/Cas): A modified version of a bacterial antiviral defense system that allows highly selective and efficient genome editing.

Counter-Screen: An assay designed to identify compounds with an undesirable mechanism of action, such as interfering with the assay technology.

Cytochrome P450 (CYP 450): A large and diverse group of enzymes that play a major role in drug metabolism.

Developability: The suitability of an antibody for manufacturing, formulation, and use in the clinic.

DMSO: A solvent often used as a vehicle in scientific experiments and assays (due to its ability to dissolve polar and nonpolar compounds while being miscible to water); also known as dimethyl sulfoxide.

Drug Master File: A confidential document submitted to the FDA (or national regulatory agency) by a company outlining specifications for the manufacturing, processing, packaging, and storing of a therapeutic agent(s).

Drug Repurposing: Finding a new indication, formulation, or route of administration for an existing drug.

Drug Target: A molecule in the body, usually a protein, that is associated with a specific pathology (disease process) that could be addressed by a drug to produce a desired therapeutic effect.

Edge-Effect: Situation in which outside wells of a multiwell plate have a bias toward different values than the rest of the plate during an assay.

Fc Domain: Tail region of an antibody that interacts with cell surface receptors.

Genome-Wide Association Study (GWAS): A method that scans the genome of a large number of subjects to identify genetic makers associated with a particular trait.

Good Laboratory Practice (GLP): Extensive documentation of each procedural step to ensure high quality, reproducible studies.

Guide RNA (gRNA): An RNA molecule that forms a complex with the CRISPR/Cas system to direct the complex to a target sequence for nucleotide editing.

High Content Screening: Cell-based HTS method that uses an automated microscope and automated image analysis to measure changes in expression or location of a macromolecule.

High-Throughput Screening (HTS): A method commonly used in drug discovery in which automated equipment is used to screen a large compound library (>100,000 compounds) against a target, in hopes of identifying compounds that produce a desired effect.

Host: A living source that allows a pathogen to live.

Human Ether-à-Go-Go-Related Gene (hERG) Channel: A potassium ion channel that is important to normal electrical activity of the heart. Inhibition of this channel can lead to sometimes fatal cardiac arrhythmias.

Humanization: Introducing mutations to a nonhuman antibody to bring its sequence closer to a natural human antibody sequence.

Immunogenicity: The capacity of a therapeutic protein to elicit an undesired immune response.

Inhibitor (competitive): Molecule that binds to the target enzyme and excludes substrate binding (and vice versa).

Inhibitor (noncompetitive): Molecule that binds to the target enzyme independent of substrate binding.

Inhibitor (uncompetitive): Molecule that binds only to the target enzyme-substrate complex.

Institutional Review Board (IRB): A committee formally designated by an institution to review, approve the initiation of, and conduct periodic reviews of biomedical research involving human subjects. The purpose of the IRB is to protect the rights and welfare of subjects/patients who are participating in the research. In most countries, this committee is called the Ethics Committee.

Investigational New Drug Application (IND): Document filed with FDA prior to initiating research on human subjects using any drug that has not been previously approved for the proposed clinical indication, dosing regimen, or patient population.

K_m (Michaelis constant): substrate concentration at which an enzymatic reaction rate is ½ of the maximal reaction rate. K_m is a way to characterize the enzyme's affinity for the substrate.

Lipophilicity: The tendency of a molecule to partition between oil and water.

Mix and Measure Assay: An assay that does not require washing away any of its components.

Miniaturization: Refers to decreasing both the volume and the plate surface area occupied by a single assay well. Common assay volumes (and surface areas) are: 100 uL (320 mm²) in 96-well plates, 40 uL (84 mm²) in 384-well plates, and 4 uL (30 mm²) in 1536-well plates.

Modeling: A simulation that captures biological relationships and either describes or predicts biological outputs.

New Drug Application (NDA): FDA documentation to obtain approval for the sales and marketing of a new drug in the US.

Off-Label: Indications not listed on the drug label (and therefore not evaluated by FDA).

Omics: A branch of science dealing with large-scale, quantifiable datasets that aim to capture a nonbiased comprehensive view of a biological system (*e.g.*, genomics, metabolomics, proteomics, transcriptomics). "Omics" have become a major source of target identification in drug discovery.

Optimization: Medicinal chemistry effort to improve the properties of a chemical lead.

Pathogen: A bacterium, virus, or other microorganism that can cause disease.

Pharmacodynamics (PD): Measurements of what a drug does to the body.

Pharmacokinetics (PK): Measurements of what the body does to a drug (*i.e.*, absorption, distribution, metabolism, and excretion [ADME]).

Polyspecificity: The ability of one antibody to bind a variety of epitopes in different antigens.

Pr(x): Probability that x will occur.

Prediction: Estimating an output measurement using input biological data.

Screening: Large-scale analysis that simultaneously measures effects for several biological molecules.

Small Interfering RNA (siRNA): A short (20–24 base-pair), double-stranded RNA molecule that acts in the RNAi pathway and is used to silence gene expression.

Tool Compound: A molecule used to establish proof-of-concept (testing whether modulating the target has the desired biological effect), but may not be suitable as a drug.

Vaccine: Substance used to stimulate the production of antibodies that provide immunity against one or several diseases.

Variable Region: Domain on an antibody responsible for antigen-binding specificity.

Woodchuck Hepatitis Virus Post-transcriptional Regulatory Element (WPRE): A DNA sequence that increases expression of genes delivered by viral vectors.

Z'-Factor: Measure of assay signal relative to background noise.

Key Abbreviations

ADME	Absorption, Distribution, Metabolism and Excretion
ADCC	Antibody-Dependent Cellular Cytotoxicity
ADCP	Antibody-Dependent Cellular Phagocytosis
ASO	Antisense Oligonucleotide
AI	Artificial Intelligence
BCR	B Cell Receptor
CIRM	California Institute for Regenerative Medicine

CDCM	Center for Definitive and Curative Medicine
CGID	Center for Genetic Immune Diseases
CAR	Chimeric Antigen Receptor
CAR T cell	Chimeric Antigen Receptor T cell
CDC	Complement-Dependent Cytotoxicity
cDNA	Complementary DNA
cGMP	Current Good Manufacturing Practice
dd	Deaminase Domain
DTA	Diagnostic Test Accuracy
dsRNA	Double-Stranded RNA
EHR	Electronic Health Records
ESC	Enhanced Stabilization Chemistry
EMA	European Medicines Agency
FDA	Food and Drug Administration
GvHD	Graft-vs-Host Disease
EC_{50}	Half Maximal Effective Concentration
IC_{50}	Half Maximal Inhibitory Concentration
HSPCT	Hematopoietic Stem and Progenitor Cell Transplantation
HSPCs	Hematopoietic Stem and Progenitor Cells
HDR	Homology Directed Repair
hERG	Human Ether-à-Go-Go-Related Gene
HLA	Human Leukocyte Antigen
IPEX	Immune Dysregulation, Polyendocrinopathy, Enteropathy, X-Linked
IRB	Institutional Review Board
IND	Investigational New Drug Application
MOA	Mechanism of Action
MMEJ	Microhomology-Mediated End Joining
miRNA	MicroRNA
mAb	Monoclonal Antibody
mRNA	Messenger RNA
NIH	National Institutes of Health
NCATS	National Center for Advancing Translational Sciences
NK	Natural Killer cells
NHEJ	Non-Homologous End Joining
NMR	Nuclear Magnetic Resonance
ODE	Ordinary Differential Equation
PS	Phosphonothioate
PoC	Proof-of-Concept
rAAV	Recombinant Adeno-Associated Virus
RNAi	RNA Interference
SCID	Severe Combined Immunodeficiency
STC	Standard Template Chemistry

SAR	Structure–Activity Relationships
TCR	T Cell Receptor
TALENs	Transcription Activator-Like Effector Nucleases
SCID-X1	X-Linked Severe Combined Immunodeficiency
ZFNs	Zinc Finger Nucleases

References

Ahn K (2017) The worldwide trend of using botanical drugs and strategies for developing global drugs. BMB Rep 50:111–116

Aiuti A, Cattaneo F, Galimberti S et al (2009) Gene therapy for immunodeficiency due to adenosine deaminase deficiency. N Engl J Med 360:447–458

Akinc A, Maier MA, Manoharan M et al (2019) The Onpattro story and the clinical translation of nanomedicines containing nucleic acid-based drugs. Nat Nanotechnol 14:1084–1087

Bates A, Power CA (2019) David vs. Goliath: the structure, function, and clinical prospects of antibody fragments. Antibodies 8:28

Borna S, Lee E, Sato Y, Bacchetta R (2022) Towards gene therapy for IPEX syndrome. Eur J Immunol 52:705–716

Carroll D (2014) Genome engineering with targetable nucleases. Annu Rev Biochem 83:409–439

Cavazzana-Calvo M, Hacein-Bey S, de Saint BG et al (2000) Gene therapy of human severe combined immunodeficiency (SCID)-X1 disease. Science 288:669–672

Cepika A-M, Sato Y, Liu JM-H, Uyeda MJ, Bacchetta R, Roncarolo MG (2018) Tregopathies: monogenic diseases resulting in regulatory T-cell deficiency. J Allergy Clin Immunol 142:1679–1695

Chames P, Van Regenmortel M, Weiss E, Baty D (2009) Therapeutic antibodies: successes, limitations and hopes for the future. Br J Pharmacol 157:220–233

Char DS, Shah NH, Magnus D (2018) Implementing machine learning in health care—addressing ethical challenges. N Engl J Med 378:981–983

Charlesworth CT, Deshpande PS, Dever DP et al (2019) Identification of preexisting adaptive immunity to Cas9 proteins in humans. Nat Med 25:249–254

Code of Federal Regulations (2023) Title 21: food and drugs, subchapter D: drugs for human use, part 312: investigational new drug application, subpart A: general provisions, Sec. 312.2: applicability. https://www.accessdata.fda.gov/scripts/cdrh/cfdocs/cfcfr/CFRSearch.cfm?fr=312.2

Copeland RA (2003) Mechanistic considerations in high-throughput screening. Anal Biochem 320:1–12

Cox DBT, Gootenberg JS, Abudayyeh OO, Franklin B, Kellner MJ, Joung J, Zhang F (2017) RNA editing with CRISPR-Cas13. Science 358:1019–1027

Cragg GM, Newman DJ (2013) Natural products: a continuing source of novel drug leads. Biochim Biophys Acta Gen Subj 1830:3670–3695

Das P, Sercu T, Wadhawan K et al (2021) Accelerated antimicrobial discovery via deep generative models and molecular dynamics simulations. Nat Biomed Eng 5:613–623

Dever DP, Bak RO, Reinisch A et al (2016) CRISPR/Cas9 β-globin gene targeting in human haematopoietic stem cells. Nature 539:384–389

Doudna JA, Charpentier E (2014) The new frontier of genome engineering with CRISPR-Cas9. Science 346:1258096

FDA-NIH Biomarker Working Group (2021) BEST (Biomarkers, EndpointS, and other Tools) resource. National Institutes of Health, Silver Spring

Ferrari S, Vavassori V, Canarutto D, Jacob A, Castiello MC, Javed AO, Genovese P (2021) Gene editing of hematopoietic stem cells: hopes and hurdles toward clinical translation. Front Genome Ed 3:618378

Ferreira LMR, Muller YD, Bluestone JA, Tang Q (2019) Next-generation regulatory T cell therapy. Nat Rev Drug Discov 18:749–769

Garg N, Schiebinger L, Jurafsky D, Zou J (2018) Word embeddings quantify 100 years of gender and ethnic stereotypes. Proc Natl Acad Sci 115:E3635–E3644

(2021) Gene therapies should be for all. Nat Med 27:1311–1311

Gironda-Martínez A, Donckele EJ, Samain F, Neri D (2021) DNA-encoded chemical libraries: a comprehensive review with succesful stories and future challenges. ACS Pharmacol Transl Sci 4:1265–1279

Godinez WJ, Hossain I, Lazic SE, Davies JW, Zhang X (2017) A multi-scale convolutional neural network for phenotyping high-content cellular images. Bioinformatics 33:2010–2019

Goldstein BA, Navar AM, Pencina MJ, Ioannidis JPA (2017) Opportunities and challenges in developing risk prediction models with electronic health records data: a systematic review. J Am Med Inform Assoc 24:198–208

Goodwin M, Lee E, Lakshmanan U, Froessl L, Barzaghi F, Passerini L, Narula M (2020a) CRISPR-based gene editing enables FOXP3 gene repair in IPEX patient cells. Sci Adv 6:eaaz0571

Goodwin M, Lee E, Lakshmanan U et al (2020b) CRISPR-based gene editing enables FOXP3 gene repair in IPEX patient cells. Sci Adv 6:eaaz0571

Gradl S, Steuber H, Weiske J et al (2021) Discovery of the SMYD3 inhibitor BAY-6035 using Thermal Shift Assay (TSA)-based high-throughput screening. SLAS Discov Adv Sci Drug Discov 26:947–960

Hendel A, Bak RO, Clark JT et al (2015) Chemically modified guide RNAs enhance CRISPR-Cas genome editing in human primary cells. Nat Biotechnol 33:985–989

Herzog RW (2010) Gene therapy for SCID-X1: round 2. Mol Ther J Am Soc Gene Ther 18:1891

High KA, Roncarolo MG (2019) Gene therapy. N Engl J Med 381:455–464

Hirakawa MP, Krishnakumar R, Timlin JA, Carney JP, Butler KS (2020) Gene editing and CRISPR in the clinic: current and future perspectives. Biosci Rep 40:BSR20200127

Ho M (2018) Inaugural editorial: searching for magic bullets. Antib Ther 1:1–5

Hogan MJ, Pardi N (2022) mRNA vaccines in the COVID-19 pandemic and beyond. Annu Rev Med 73:17–39

Hsiao WLW, Liu L (2010) The role of traditional Chinese herbal medicines in cancer therapy – from TCM theory to mechanistic insights. Planta Med 76:1118–1131

Hu B, Zhong L, Weng Y, Peng L, Huang Y, Zhao Y, Liang X-J (2020) Therapeutic siRNA: state of the art. Signal Transduct Target Ther 5:1–25

Huang R, Southall N, Wang Y, Yasgar A, Shinn P, Jadhav A, Nguyen D-T, Austin CP (2011) The NCGC Pharmaceutical Collection: a comprehensive resource of clinically approved drugs enabling repurposing and chemical genomics. Sci Transl Med 3:80ps16

Iglesias-Lopez C, Agustí A, Vallano A, Obach M (2021) Current landscape of clinical development and approval of advanced therapies. Mol Ther Methods Clin Dev 23:606–618

Jain T, Sun T, Durand S et al (2017) Biophysical properties of the clinical-stage antibody landscape. Proc Natl Acad Sci 114:944–949

Ji H, Li X, Zhang H (2009) Natural products and drug discovery: can thousands of years of ancient medical knowledge lead us to new and powerful drug combinations in the fight against cancer and dementia? EMBO Rep 10:194–200

Jinek M, Chylinski K, Fonfara I, Hauer M, Doudna JA, Charpentier E (2012) A programmable dual-RNA–guided DNA endonuclease in adaptive bacterial immunity. Science 337:816–821

June CH, O'Connor RS, Kawalekar OU, Ghassemi S, Milone MC (2018) CAR T cell immunotherapy for human cancer. Science 359:1361–1365

Katrekar D, Yen J, Xiang Y, Saha A, Meluzzi D, Savva Y, Mali P (2022) Efficient in vitro and in vivo RNA editing via recruitment of endogenous ADARs using circular guide RNAs. Nat Biotechnol 40:1–8

Koch R (1890) Verber Bakteriologische Forschung. In: Verhandlungen des X Internationalen Medizinischen Kongresses, pp 35–74

Kohn MA (2022a) Studies of diagnostic test accuracy: partial verification bias and test result-based sampling. J Clin Epidemiol 145:179–182

Kohn MA (2022b) Key concepts in clinical epidemiology: reporting on the accuracy of continuous tests. J Clin Epidemiol 141:157–160

Kohn MA, Carpenter CR, Newman TB (2013) Understanding the direction of bias in studies of diagnostic test accuracy. Acad Emerg Med 20:1194–1206

Laustsen AH, Greiff V, Karatt-Vellatt A, Muyldermans S, Jenkins TP (2021) Animal immunization, in vitro display technologies, and machine learning for antibody discovery. Trends Biotechnol 39:1263–1273

Li JW-H, Vederas JC (2009) Drug discovery and natural products: end of an era or an endless frontier? Science 325:161–165

Lipinski CA, Lombardo F, Dominy BW, Feeney PJ (1997) Experimental and computational approaches to estimate solubility and permeability in drug discovery and development settings. Adv Drug Deliv Rev 23:3–25

Lord SJ, Irwig L, Simes RJ (2006) When is measuring sensitivity and specificity sufficient to evaluate a diagnostic test, and when do we need randomized trials? Ann Intern Med 144:850–855

Majzner RG, Mackall CL (2018) Tumor antigen escape from CAR T-cell therapy. Cancer Discov 8:1219–1226

Maraganore J (2022) Reflections on Alnylam. Nat Biotechnol 40:1–10

Masiuk KE, Laborada J, Roncarolo MG, Hollis RP, Kohn DB (2019) Lentiviral gene therapy in HSCs restores lineage-specific Foxp3 expression and suppresses autoimmunity in a mouse model of IPEX syndrome. Cell Stem Cell 24:309–317.e7

Merkle T, Merz S, Reautschnig P, Blaha A, Li Q, Vogel P, Wettengel J, Li JB, Stafforst T (2019) Precise RNA editing by recruiting endogenous ADARs with antisense oligonucleotides. Nat Biotechnol 37:133–138

Monian P, Shivalila C, Lu G et al (2022) Endogenous ADAR-mediated RNA editing in non-human primates using stereopure chemically modified oligonucleotides. Nat Biotechnol 40:1–10

Montiel-Gonzalez MF, Vallecillo-Viejo I, Yudowski GA, Rosenthal JJC (2013) Correction of mutations within the cystic fibrosis transmembrane conductance regulator by site-directed RNA editing. Proc Natl Acad Sci 110:18285–18290

Mullard A (2021) FDA approves 100th monoclonal antibody product. Nat Rev Drug Discov 20:491–495

Murray CW, Rees DC (2009) The rise of fragment-based drug discovery. Nat Chem 1:187–192

Naldini L (2019) Genetic engineering of hematopoiesis: current stage of clinical translation and future perspectives. EMBO Mol Med 11:e9958

National Center for Advancing Translational Sciences (2022) About new therapeutic uses. https://ncats.nih.gov/ntu/about

Newman TB, Kohn MA (2019) Chapter 3: multi-level and continuous tests. In: Evidence-based diagnosis. An introduction to clinical epidemiology, 2nd edn. Cambridge University Press, Cambridge/New York, pp 47–74

Nguengang Wakap S, Lambert DM, Olry A, Rodwell C, Gueydan C, Lanneau V, Murphy D, Le Cam Y, Rath A (2020) Estimating cumulative point prevalence of rare diseases: analysis of the Orphanet database. Eur J Hum Genet EJHG 28:165–173

Passerini L, Rossi Mel E, Sartirana C, Fousteri G, Bondanza A, Naldini L, Roncarolo MG, Bacchetta R (2013) CD4$^+$ T cells from IPEX patients convert into functional and stable regulatory T cells by FOXP3 gene transfer. Sci Transl Med 5:215ra174

Pavel-Dinu M, Wiebking V, Dejene BT et al (2019) Gene correction for SCID-X1 in long-term hematopoietic stem cells. Nat Commun 10:1634

Porteus MH (2019) A new class of medicines through DNA editing. N Engl J Med 380:947–959

Porteus MH, Baltimore D (2003) Chimeric nucleases stimulate gene targeting in human cells. Science 300:763

Prudent R, Annis DA, Dandliker PJ, Ortholand J-Y, Roche D (2021) Exploring new targets and chemical space with affinity selection-mass spectrometry. Nat Rev Chem 5:62–71

Qasim W, Zhan H, Samarasinghe S, Adams S, Amrolia P, Stafford S, Butler K, Rivat C (2017) Molecular remission of infant B-ALL after infusion of universal TALEN gene-edited CAR T cells. Sci Transl Med 9:eaaj2013

Qu L, Yi Z, Zhu S et al (2019) Programmable RNA editing by recruiting endogenous ADAR using engineered RNAs. Nat Biotechnol 37:1059–1069

Rajkomar A, Oren E, Chen K et al (2018) Scalable and accurate deep learning with electronic health records. Npj Digit Med 1:1–10

Reautschnig P, Wahn N, Wettengel J et al (2022) CLUSTER guide RNAs enable precise and efficient RNA editing with endogenous ADAR enzymes in vivo. Nat Biotechnol 40:1–10

Roth TL, Puig-Saus C, Yu R et al (2018) Reprogramming human T cell function and specificity with non-viral genome targeting. Nature 559:405–409

Saah AJ, Hoover DR (1997) "Sensitivity" and "specificity" reconsidered: the meaning of these terms in analytical and diagnostic settings. Ann Intern Med 126:91–94

Santoni de Sio FR, Passerini L, Valente MM, Russo F, Naldini L, Roncarolo MG, Bacchetta R (2017) Ectopic FOXP3 expression preserves primitive features of human hematopoietic stem cells while impairing functional T cell differentiation. Sci Rep 7:15820

Sato Y, Passerini L, Piening BD, Uyeda MJ, Goodwin M, Gregori S, Snyder MP, Bertaina A, Roncarolo M-G, Bacchetta R (2020) Human-engineered Treg-like cells suppress FOXP3-deficient T cells but preserve adaptive immune responses in vivo. Clin Transl Immunol 9:e1214

Sayitoglu EC, Freeborn RA, Roncarolo MG (2021) The Yin and Yang of Type 1 regulatory T cells: from discovery to clinical application. Front Immunol 12:693105

Smith AJ (2015) New horizons in therapeutic antibody discovery: opportunities and challenges versus small-molecule therapeutics. J Biomol Screen 20:437–453

Stafforst T, Schneider MF (2012) An RNA–deaminase conjugate selectively repairs point mutations. Angew Chem Int Ed 51:11166–11169

Stumpfe D, Bajorath J (2020) Current trends, overlooked issues, and unmet challenges in virtual screening. J Chem Inf Model 60:4112–4115

Torng W, Altman RB (2019) Graph convolutional neural networks for predicting drug-target interactions. J Chem Inf Model 59:4131–4149

U.S. Food and Drug Administration. Center for Drug Evaluation and Research (CDER). Botanical Drug Development: Guidance for Industry (2016) Silver Spring. Available at: https://www.fda.gov/media/93113/download

Vogel P, Moschref M, Li Q, Merkle T, Selvasaravanan KD, Li JB, Stafforst T (2018) Efficient and precise editing of endogenous transcripts with SNAP-tagged ADARs. Nat Methods 15:535–538

Wilson JL, Gravina A, Grimes K (2022) From random to predictive: a context-specific interaction framework improves selection of drug protein–protein interactions for unknown drug pathways. Integr Biol 14:13–24

Wu G, Yuan Y, Hodge CN (2003) Determining appropriate substrate conversion for enzymatic assays in high-throughput screening. J Biomol Screen 8:694–700

Wu C, Lee S-L, Taylor C, Li J, Chan Y-M, Agarwal R, Temple R, Throckmorton D, Tyner K (2020) Scientific and regulatory approach to botanical drug development: a U.S. FDA Perspective J Nat Prod 83:552–562

Yang B, Mallett S, Takwoingi Y et al (2021) QUADAS-C: a tool for assessing risk of bias in comparative diagnostic accuracy studies. Ann Intern Med 174:1592–1599

Yi Z, Qu L, Tang H et al (2022) Engineered circular ADAR-recruiting RNAs increase the efficiency and fidelity of RNA editing in vitro and in vivo. Nat Biotechnol:1–10

Yuan B, Shen C, Luna A, Korkut A, Marks DS, Ingraham J, Sander C (2021) CellBox: interpretable machine learning for perturbation biology with application to the design of cancer combination therapy. Cell Syst 12:128–140.e4

Zhang J-H, Chung TDY, Oldenburg KR (1999) A simple statistical parameter for use in evaluation and validation of high throughput screening assays. J Biomol Screen 4:67–73

On the Way to the Clinic

3

Daria Mochly-Rosen, Kevin Grimes, C. Glenn Begley,
Dirk Mendel, Werner Rubas, Emily Egeler,
Collen Masimirembwa, Terrence F. Blaschke,
and Michael Taylor

Daria Mochly-Rosen, Ed, Kevin Grimes, Ed.

Once a "hit" drug has been identified, it is important to evaluate it in appropriate animal models for the indicated pathology and to apply the highest standards of research to ensure the integrity of the findings. When performing drug discovery

D. Mochly-Rosen (✉) · K. Grimes
Chemical and Systems Biology, Stanford University School of Medicine, Stanford, CA, US
e-mail: sparkmed@stanford.edu; kgrimes@stanford.edu

D. Mendel
SPARK at Stanford, Stanford University School of Medicine, Stanford, CA, US

C. G. Begley
SPARK at Stanford Advisor, Stanford, CA, US

E. Egeler · T. F. Blaschke
Stanford University School of Medicine, Stanford, CA, US

W. Rubas
Sutro Biopharma, Inc., South San Francisco, CA, US

C. Masimirembwa
African Institute of Biomedical Science and Technology, Harare, Zimbabwe

M. Taylor
NonClinical Safety Assessment, Mountain View, CA, US

© The Author(s), under exclusive license to Springer Nature Switzerland AG 2023
D. Mochly-Rosen, K. Grimes (eds.), *A Practical Guide to Drug Development in Academia*, https://doi.org/10.1007/978-3-031-34724-5_3

research, we must be particularly attentive to the robustness of our experiments, because inability to reproduce academic data continues to be a sticking point when projects are transferred to industry. Our experiments must be appropriately blinded, statistically powered, and meticulously documented so that our findings are worthy of the large investment required for further translation into a drug.

This chapter walks through the essential preclinical drug development steps to ensure that the compound has optimal pharmacological features. These features include understanding how the drug is working on the body (pharmacodynamics; PD) and what the body does to the drug (the drug's absorption, distribution, metabolism, and excretion, a field collectively termed pharmacokinetics; PK). We also must remember that whereas animal models are often genetically identical, the great genetic variation among humans must be considered when developing a drug. Genetic variations in the molecular target of the drug—its PD, as well as in enzymes and proteins that affect its PK, are critical to evaluate.

Early evaluation of these characteristics will increase the probability that clinical trials will succeed. If a subset of patients has a protein target that is not recognized by the drug, their inclusion in the trial will reduce the overall benefit of the tested drug. Similarly, genetic variations in drug absorption, metabolism, and/or secretion may result in lack of response or unacceptable side effects of the tested drug. Unfortunately, these latter features are often discovered after drug approval, yet patient suffering could have been readily prevented by introducing pharmacogenomics studies during drug development. Drug formulation and route of administration must be established to maximize treatment efficacy and ease of use for the patient and physician. Lastly, preclinical safety studies must be conducted to predict potential drug toxicities and establish safe dosing ranges.

These topics are rarely addressed in traditional academic drug discovery efforts. We hope that you will incorporate them into your research plan, because without such considerations, even a great drug is likely to fail.

3.1   When to Begin Animal Studies

Daria Mochly-Rosen

We have identified a new chemical entity or a known drug that affects our validated target/pathway and has shown its efficacy in a cell-based assay. What is the next step?

Experts are divided on whether it is advisable to begin animal studies right away or whether it is better to first identify the optimal compound. On one hand, by generating and testing analogs of the original "hit," it may be possible to improve potency or specificity for the target. In vitro studies to obtain an optimal formulation for a drug or simply increase solubility can also improve the chance for success once animal studies begin. There are other considerations, such as in vitro assessment of drug toxicity and metabolism, including liver enzyme assays, human ether-à-go-go-related gene (hERG) channel effects, etc. In other words, we can easily spend a year and thousands of dollars in studies aimed at improving our initial hit.

On the other hand, in vitro and cell-based assays are usually cheaper and faster to run than animal studies, but they are not always predictive of the in vivo behavior of the molecule—which is ultimately most important for determining if our hit will make a good drug. So, how are drug development programs to decide, with their limited funds, between screening many analogs in vitro versus testing only a handful of molecules in animals? As an academic that has followed over 230 programs within SPARK at Stanford, my answer to this question is simple.

3.1.1   Take a Shortcut

We should start animal studies as soon as we can. It is true that many improvements to our compound can be made, but a short in vivo study can be extremely valuable in helping to optimize the compound and induce greater interest from partners and investors. A great deal can be learned from an imperfect drug. We might even be lucky and find that our compound shows a therapeutic benefit and drug-like properties!

We must also recognize that failure to demonstrate efficacy at this stage is not a reason to discontinue our project. These are exploratory studies, and much can still be done to improve the compound's selectivity, potency, solubility, bioavailability, safety, metabolism, route of administration, and final formulation. The "take a shortcut" approach is illustrated in Fig. 3.1, with key concepts further discussed in this section.

3.1.2   What Animal Model to Use?

It is best to read the literature and use an animal model that is accepted in the field for the given indication. It is inadvisable to develop a new model for this first in vivo trial. Better yet, we can find a collaborator that is using this animal model and have them do the study for us. It is rare that such a study will generate new intellectual

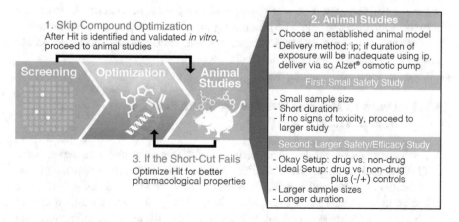

Fig. 3.1 When to begin animal studies. Abbreviations: *ip* Intraperitoneal, *sc* Subcutaneous

property—and the collaborator can provide an independent and unbiased assessment of our compound. It is worth noting that the animal model used for safety studies may not be the same model used for efficacy studies (for more on animal models, refer to Sect. 3.3.3). Additionally, despite limited financial resources, we must make sure that our animal studies are robust enough. Review Sect. 3.2, Robustness of Preclinical Studies, before embarking on this important part of your program.

3.1.3 How to Deliver the Drug?

Even if we believe that oral administration is the ideal route for our clinical indication, it is ill-advised to attempt to do the first efficacy study in animals using oral gavage. Instead, consider intraperitoneal (ip) injection. If the drug is not very soluble, we can deliver the drug with ethanol, DMSO, or polyethylene glycol; animals will tolerate quite a high dose of these solvents. Consult your animal facility for what is well tolerated in your animal species. If there is concern that the drug dose will be too low or exposure too short using ip injection, consider using a subcutaneous (sc) Alzet® osmotic pump. The company's website details various sizes and recommended solvents as well as training on how to implant them.

> **Box 3.1: What Surprised an Academic?**
> When selecting a delivery formulation for these initial animal studies, simpler is better. We once used an over the counter beauty lotion as the vehicle for an initial topical delivery study because it had the desired aqueous formulation properties. —*DM-R*

3.1.4 Start with a Small Safety Study

To make sure that the drug dose is not fatal, inject a couple of healthy animals and observe them for a few hours or days for obvious signs of toxicity. A veterinary nurse or technician can help with monitoring for adverse events. Once we know that the dose selected is not acutely toxic, we can jump into efficacy and longer safety studies in the chosen animal model of disease.

3.1.5 Learn as Much as You Can from the First (Pilot) In Vivo Study

Animals are precious and should be used sparingly. Therefore, plan your experiments carefully to include proper controls. For a first pilot study, it is okay to just compare drug-treated to nontreated animals. However, it is better to also include a drug that is known to be efficacious in the model—if one is available—as a positive control and a vehicle as a negative control. When euthanizing the animals, collect as many organs and bio-fluids as possible for analysis. A pathologist can advise how to preserve the tissues and store samples for later analysis. We should attempt to collect as much data as possible relevant to our disease and to compound safety. The results will be most helpful if drug concentrations in the blood (and other relevant tissues) can be determined for each dose administered. The bottom line: We need to maximize the information obtained from this first set of studies.

3.1.6   If the Shortcut Failed

We are not done! Remember that we have committed to take the long route even if the shortcut failed. We can go back and perform further structure–activity relationship (SAR) studies with analogs of our hit and additional studies on drug solubility and in vitro toxicity. We can now focus on correcting the problems identified based upon the first in vivo experiment.

3.1.7   If the Shortcut Succeeded

Congratulations! The work has just begun. Now we have more compelling data that the project is worth pursuing. Make sure to consult Sect. 3.2 on robust preclinical work and Sect. 3.3 on in vivo pharmacology to plan your next steps.

Later in this chapter, you will learn that exploring pharmacokinetics (PK) prior to efficacy studies has many advantages. In SPARK, we provide multiple, and sometimes what appears to be opposing opinions, as there is no single path for drug development. After reading both sections, it is up to you to decide whether you will perform efficacy studies first (as described above), or whether you will conduct proper PK studies prior to efficacy studies; both eventually need to be completed. However, based on the resources available to you, you will need to choose the order best suited for your project.

> **The Bottom Line**
> An early small in vivo study can be extremely helpful in demonstrating both efficacy and preliminary toxicity of the drug. Results can also inform further rounds of optimization of the compound. During initial animal studies, the drug should generally be administered using a parenteral route (ip or sc via an osmotic infusion pump).

3.2   Robustness of Preclinical Studies

Daria Mochly-Rosen and C. Glenn Begley

A number of publications challenge the robustness of academic preclinical studies. In one report, only 11% of published "positive" preclinical cancer studies from academic labs could be reproduced by the original researchers when working with

Amgen scientists. Most academic scientists were unable to reproduce their own published work, even when working in their own laboratories with their own reagents, when they were blinded to the experimental groups (Begley and Ellis 2012). In another report, Bayer scientists found that ~75% of published academic studies could not be replicated, which resulted in termination of efforts to develop therapeutics based on these academic findings (Prinz et al. 2011).

These findings were recently confirmed in a prospective, thorough analysis that showed only 13% of "positive" preclinical cancer findings could be reproduced (Errington et al. 2021a). The irreproducibility of the majority of academic studies, even when published in "high-profile" journals, has been long known to industry. As a result, it is routine practice for investors to attempt to independently reproduce key findings in another lab prior to any investment. So why is it that published findings are of such poor quality?

The following discussion focuses on academic data related to animal studies. Issues that won't be discussed include, for example, the importance of using the right animal models, confirmation using patient specimens, and use of appropriate study "endpoints." These issues are all discussed in later sections of this chapter.

Instead, irreproducible studies typically reflect a failure to perform studies using a solid methodology—factors that can be easily remedied by the academic researchers (Fig. 3.2).

Box 3.2: What Surprised an Academic?

In 2004, I temporarily moved from my academic lab to serve as the Chief Scientific Officer (CSO) of KAI Pharmaceuticals. I was hurt when our Chief Executive Officer (CEO), who holds a BA in history, told me, "You will now learn that your academic work is not as robust as industry's standard." Like you, I take great pride in our work in academia. I felt that conducting blinded studies, using several species, and reproducing the work in independent labs all combined to ensure high quality and valid data. That was not enough, I quickly learned. —DM-R

3.2.1 Factors that Contribute to Irreproducible Data

3.2.1.1 Failure to Focus on Methods Drives Irreproducible Results

While animal studies can be affected by factors over which researchers may have little control (as briefly discussed below), the principal problem lies with the inadequate methodology employed by the researchers themselves.

The areas where researchers may have less control include, for example, the nocturnal nature of rats. Therefore, data related to their immune response, eating, exercise, ability to learn tasks, etc., may be impacted by the time of day when the experiment is conducted. Equally, chow feed may be a variable that can affect

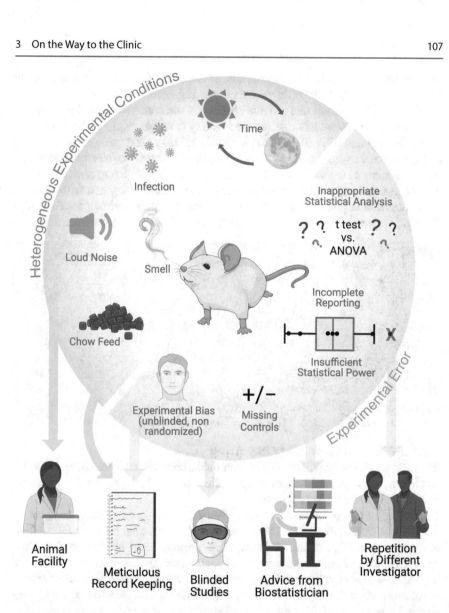

Fig. 3.2 Factors that contribute to irreproducible data in academia and solutions to mitigate these errors. Abbreviations: *ANOVA* Analysis of Variance. (Created with BioRender.com)

animal-derived data; some feed is rich in soy and may therefore contribute feminizing hormones to both males and females; variation in the feed may affect response to drug uptake and metabolism.

Other potentially confounding factors relate to the housing conditions, including noise, strong smells, and crowding. Latent or full-blown infection by viruses, bacteria, mites, and other parasites can also impact study results (see Box 3.3). All these variables should be minimized where possible and detailed information should be recorded and provided by researchers in any publication or report.

Although these issues can provide challenges when attempting to interpret discordant results between laboratories, the inability to repeat results in the same laboratory (Begley and Ellis 2012; Errington et al. 2021a) speaks to more fundamental and important issues including lack of blinding of researchers, lack of randomization of animals to the various treatment groups, inadequate cohort sizes, failure to include positive and negative controls, use of nonvalidated reagents, failure to link PK exposure and pharmacodynamics (PD), lack of a prespecified power calculation, and inappropriate analysis of results (Begley 2013; Errington et al. 2021b).

Box 3.3: Unsuspected Confounding Factors Can Affect Reproducibility
Lack of reproducibility of preclinical reports does not necessarily mean that the data are fabricated or wrong (although that may be the case). One of the better-documented cases of inability to reproduce data in mice relates to the induction of type I diabetes in non-obese diabetic (NOD) mice. Initial claims attributed increased diabetes incidence reported by some groups to the difference in housing the mice in germ-free conditions. However, more recent data showed that intestinal microbiota are the critical confounding factor; presence of Bacillus cereus in the gut delayed onset and reduced incidence of type 1 diabetes (King and Sarvetnick 2011).

3.2.1.2 Bias and Incomplete Reporting

It is critical that investigators who assess animal data are blinded to the experimental conditions; unintended bias can greatly affect the analysis, especially when the endpoint determinants are subjective.

Another bias problem results from dismissing and not reporting negative or inconsistent data. Although the investigator may have a strong rationale for wanting to exclude data from certain animals, the reasoning must be described and the criteria established before the experiment commences and completely described in the Methods section of the published report. This allows readers to draw their own conclusions about the legitimacy of the approach. All data (positive and negative) should be reported, as they may help identify important variables to consider in human studies. For example, the observation that sex and age can affect the therapeutic response to drugs in animal models of heart attack was not reported for a long time. When these findings were finally reported, reviewers started requesting that preclinical studies include animals from both sexes.

All studies should include both positive and negative controls. For example, including a group of animals treated with a drug that was previously approved for the indication will not only enable a side-by-side comparison of the benefit of our intervention, but will also confirm that the disease model is relevant. This positive control will serve as the "gold standard" in a head-to-head study and will demonstrate that the new approach has the potential to be superior (or not) to existing therapies.

This is referred to as "differentiation" from current therapeutic approaches and is essential for attracting investment to develop a new therapy.

Academics sometimes assume that certain controls are wasteful. "We have done these controls before" or "we did not have the funds to include this control again" are some of the reasons often used. However, control experiments need to be done side-by-side with the treatment arm, as unexpected factors can contribute to the outcomes. An investigator in SPARK said that they omitted an oral gavage of their control subjects before the last blood draw, only to discover later that gavage alone increases neutrophil number in the blood—possibly due to animal stress. Unfortunately, the entire study had to be repeated. Concurrent controls are always essential.

It is important that critical experiments are repeated by a different investigator in the same lab to ensure that the experimental protocol is detailed enough to be reproduced by an unbiased researcher. When Dr. Mochly-Rosen first reported on the benefit of treating animals with an inhibitor of protein kinase C delta after inducing a heart attack, the degree of improvement was so surprising that one skeptic refused to believe the results. It was good to be able to answer that three members of the lab had reproduced the same data. It was even better to report that two other labs had reproduced the data, and it was really a coup when that skeptic obtained the same data in their own laboratory.

3.2.1.3 Insufficient Statistical Power of the Study or Inappropriate Statistical Analysis

To contain costs, researchers in academia often use too few animals per treatment cohort. Unfortunately, a p-value smaller than 0.05, although deemed statistically significant, is not sufficiently robust if the study was performed, for example, with only 5 animals per treatment group. Rather than contact a statistician prior to designing a study, most researchers only contact a statistician when attempting to analyze the data from a suboptimally powered study.

Multiple commentaries urge academics to recognize the critical contribution of statisticians in preclinical research (e.g., Peers et al. 2012). Statisticians should be engaged early during study planning to ensure that the number of animals included is sufficient and that the study is powered to provide an unequivocal answer. This will not only ease the review process, but importantly will increase the rigor of the study. If the budget dictates the number of animals per group, researchers risk generating data that will be uninterpretable.

Biostatisticians can also advise on the appropriateness of the statistical tests used to analyze the results. Often, there is more than one statistical test available to compare groups, but characteristics of the data (e.g., size, distribution) may make some tests inappropriate. For example, we should not use a t-test on nonparametric data.

Box 3.4: What Surprised an Academic?
After demonstrating that our drug successfully lowered parathyroid hormone in preclinical models of secondary essential hyperparathyroidism (a common pathology in kidney dialysis patients) in three species, KAI investigators were devastated by the lack of ANY effect when the drug was administered to pigs at the request of our investors. Before giving up though, Dr. Jim Tomlinson (a superb KAI scientist), compared the sequence of the calcium sensor receptor in different species. He found that only the pig receptor is missing a cysteine residue in the putative binding site. This discovery was not only critical in explaining the discrepancy between pigs and many other species including humans, but also identified the molecular basis for the drug's action. It saved a program that resulted in an FDA-approved drug. Think like a scientist before giving up!—*DM-R*

3.2.2   Conclusions

Given the disappointing results of published studies on reproducibility, should academic groups even attempt to do preclinical studies? Drug development is a partnership between academia and industry, and academic research provides essential fuel for new drug development. In an analysis of 252 drugs approved by the Center for Drug Evaluation and Research (CDER) between 1998 and 2007, only 47% were considered scientifically novel; and academic discoveries contributed to a third of those novel molecules (Kneller 2010). In addition, almost 50% of drugs approved for orphan indications during that period were based on academic discoveries. Therefore, academic research is an important and essential engine for innovation in drug discovery. Nevertheless, as Begley and Ellis conclude, the bar for reproducibility in performing and presenting preclinical studies must be raised. More rigorous preclinical research in academia will reduce waste in both academia and industry, thus leading to less costly drug discovery efforts and greater benefit to patients.

The Bottom Line
The bar for reproducibility in performing and presenting preclinical studies carried out by academic scientists must be raised, lest innovative academic work go unnoticed by industry partners.

3.3   In Vivo Pharmacology: Multiple Roles in Drug Discovery

Dirk Mendel

In classical drug discovery, in vivo pharmacology—studying the interaction between a compound and a body—was often an early step in the drug discovery/

Recommendations
**Improve Robustness of Preclinical Studies (Expanded from Ref.
(Begley 2013))**

1. Keep detailed information about the experimental conditions.
2. Keep detailed information on the source of all reagents and lot numbers used in the study.
3. Seek the advice of statisticians to perform a power calculation to determine the minimum cohort size before commencing the study.
4. Work with statisticians during and at completion of the study to ensure that appropriate statistical tests are applied.
5. Include appropriate negative controls and—when possible—positive controls for the study.
6. Have each study reproduced by another blinded investigator in the lab, and in an independent lab if feasible.
7. Investigators should be blinded to the identity of the control and treatment groups during treatment and during data analysis.
8. Provide information on all animals that were included in the study, those that were excluded from the study, and the reasons for the exclusion.
9. Validate reagents for the intended application (e.g., selectivity of small molecule, appropriate antibody for immunohistochemistry).

development funnel and was sometimes used as one of the first "assays" to determine whether compounds exhibited evidence of the desired biological activity. However, times have changed, and in vivo pharmacology in industry has moved to later in the process, with compounds often not tested in animal models until they are reasonably optimized for potency, selectivity, and at least "acceptable" pharmacokinetic properties. As noted in Sect. 3.1, there is dissent regarding how "optimized" compounds should be before you run at least a simple animal study to look for any evidence of activity, but the primary reason for this change in strategy is that animal studies are quite expensive due to the cost of animals (especially genetically engineered animals used in disease models); the quantity of compound needed for repeat dose animal studies; the human resources necessary to maintain the animals and run the experiments, especially if the readout is efficacy with repeated daily dosing of a week or more in an animal disease model; and the growing appreciation of the ethical imperative to minimize needless use of animals in biological research. In addition, since we now typically use a target-based approach to select and conduct projects, compounds can be progressed much further along the optimization pathway using cell-based assays, including some sophisticated patient-derived cell lines, mixed cell culture systems, or even organotypic models that reasonably model more complex physiological interactions.

In light of these advancements, what is the value and objective of in vivo pharmacology studies in modern day drug discovery/development? Few would argue that the ultimate goal of in vivo pharmacology studies is to demonstrate that the compound you plan to take into clinical development can show benefit in an animal model of the human disease for which it is intended to be developed. However, a carefully thought-out in vivo pharmacology strategy can provide much more critical information and can be executed efficiently and effectively. The goal of this section is to discuss the most useful information one should try to generate since a lot of time, money, and resources are invested into these experiments.

> **Box 3.5: What Surprised an Academic?**
> The regulatory requirements for pharmacology studies in support of an investigational new drug application (IND) are quite minimal. While at least some data are expected to suggest the compound is likely to have clinical benefit, the primary concern during regulatory review of an IND application is ensuring the safety of the human subjects. Consequently, you should not design your pharmacology studies to "convince" FDA your compound is going to work; the real target audience for in vivo pharmacology studies should be the decision makers (e.g., yourself, potential investors, your management), and the goal should be to convince them that the project/molecule warrants further investment. If you can do that, FDA should have no issue with the pharmacology information supporting your IND application. —*DM*

3.3.1 Critical Questions to Consider When Designing In Vivo Pharmacology Studies

The two most critical questions addressed by in vivo pharmacology studies are:

1. **Does the compound hit the intended molecular target and have the desired biological effect in vivo?** This question is addressed by demonstrating that the compound has a target-mediated PD effect, preferably but not necessarily in the target tissue of interest. The goals are to understand the relationship between drug levels (exposure, or PK) and activity (PD); what, if any, is the time lag between drug–target interaction and PD readout; and the extent to which the target needs to be inhibited or activated to get the desired level of downstream biological activity.
2. **Does hitting the intended target have a "clinical" benefit (efficacy) in a disease-relevant animal model?** This question is typically addressed by conducting dose–response efficacy studies in the disease-relevant animal model, though in these studies you will need to include some PK sampling to make sure the drug exposure increases as the delivered dose is increased.

When addressing these two critical questions in animal studies prior to progressing compounds into clinical development, there are two philosophies regarding what the first animal study should be. Some people think it should be the dose–response study to look for evidence of efficacy in a disease-relevant model since that is what you care most about, and if you know the compound works, you can then go back and show it hits the intended target. Others, including me, think it should be a simple study to look for evidence of target-mediated PD activity.

Starting with the PD study is simpler, quicker, and cheaper and requires less compound, so you can easily do more of them. Also, you are more likely to get positive results, and the information generated tends to be very informative. In particular, the information these studies provide is critical in guiding the design of the dose–response efficacy study, which is a major investment and runs the risk of providing limited or no information if you do not choose the right dose range or dosing regimen.

A good PD study should incorporate PK sampling so that plasma drug levels can be correlated with PD response, assuming that the PD response is relatively quick (i.e., such as a kinase response rather than a change in protein levels). Extremely useful data can be obtained from a limited number of experiments conducted in animals administered a single dose of compound at 2 or 3 dose levels (Mendel et al. 2015; Fell et al. 2015). Even a single experiment with 20–30 animals, easily completed over 1 or 2 days, can provide a relatively robust understanding of the PK/PD relationship for the compound's effect on your target.

In addition to demonstrating that the compound has the desired on-target activity in vivo, the PK/PD data combined with an understanding of the amount of target modulation required to obtain the desired biological activity will increase the likelihood of testing relevant doses in subsequent efficacy studies. Without this information, you would randomly select 3 or 4 doses for the animal efficacy studies plus a vehicle control, with 10–12 animals per group, for 2 weeks or more, and may end up with little to no information, having used extensive resources and compound. You may wonder whether your drug is not efficacious or the dose is not high enough and will have to repeat the experiment at higher doses, again with no real guidance.

3.3.2   Information to Help Guide Dosing in the Clinic

While the two questions cited above are the most critical, another important goal of in vivo pharmacology in drug development is to provide some dose guidance for human studies. Most often, dose guidance comes in the form of calculating a human equivalent dose (HED) for the fully efficacious dose identified from the dose–response efficacy studies in the disease-relevant animal model. For small molecules, determination of the HED is based on standard tables for allometric scaling from animal species to humans. The mg/kg dose in the test species is converted to an equivalent mg/kg dose for humans based on body surface area, which then can be converted to a total (mg) dose for a human of standard size (typically 60–70 kg) (U.S. Food and Drug Administration 2005).

Although commonly used, this approach's shortcoming is that these calculations are designed to identify a human dose that will produce a similar exposure, or area under the plasma drug concentration vs. time curve (AUC), in humans as in the preclinical animal. This information is a way to standardize for safety, not efficacy.

While total exposure (AUC) may be the same, the shape of the PK curve typically flattens and extends when moving up in species, so curves that have the same AUC can look very different (Box 3.6). Since biological activity, which leads to efficacy, is related to extent and duration of activity on the target rather than AUC, different shaped curves can have very different levels of target engagement.

A more meaningful, and arguably, a more translatable way to provide dose guidance to the clinical group would be to provide data on activity when plasma concentration of the drug was maintained for a certain amount of time: A plasma level of X ng/mL was required to get the needed level of target engagement in the animal model(s); evidence of efficacy (i.e., the lowest efficacious dose) was obtained when this concentration of drug was maintained for Y hours; and maximum efficacy (i.e., the fully efficacious dose) was observed when this concentration of drug was maintained for Z hours. This very specific information provides clear and easily translatable guidance for clinical studies, especially when differences in plasma protein binding are considered to normalize for free drug.

Box 3.6: Shortcoming in the Typical Method to Calculate HED

Allometric scaling to calculate a HED from animal data is designed to identify a dose that will give the same total exposure (AUC) in various species. However, when moving from rodents to dogs to humans it is not unusual to see a progression of PK profiles that flatten when moving up in species. (As shown below, mouse PK may look like the curve on the left, dog like the curve in the middle, and human like the curve on the right.) While these 3 curves have the same total exposure (AUC), they do not have the same amount of time above the plasma concentration required to achieve target-mediated activity (indicated by horizonal dotted line). The HED calculated this way will not produce the same level of efficacy as that observed in the mouse efficacy study. Potential species differences in bioavailability are also not accounted for in the HED calculated using this approach.

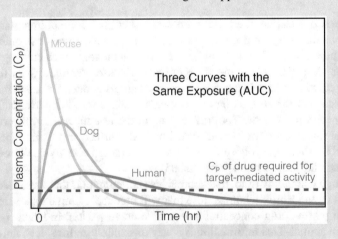

3.3.3 Selection of Animal Models

George Box, an English statistician, is credited with several versions of the statement "All models are wrong, but some are useful." He was discussing mathematical models, but his key point is equally valid for biological questions: If the model is relevant to the specific question you are asking, it can be used to address that question. We have developed some very sophisticated animal models of many human diseases, and there is no doubt that these are important for evaluating potential drug efficacy. However, it is critical to use the right model for the specific question being addressed.

When attempting to understand the PK/PD relationship for your target, the simplest, most robust system will likely be normal tissue or bodily fluid in normal animals, or cells that have been engineered to express the target in functional form, which can be studied in vitro or implanted into animals to be studied in vivo. These models can help address questions such as how much drug is required to inhibit/activate the target; how much inhibition/activation of the target is required to get the desired biological activity; and how long that level of biological activity should be maintained to have the desired downstream effect, which in the right disease-relevant model translates to efficacy.

Which animal models to use for efficacy studies is a much tougher question, and that is where biology or disease subject matter experts prove their worth. Models can appear to be very similar but can differ fundamentally regarding the target of interest, or may seem to test clinically relevant activity but instead assay something quite different. The clinical success or failure of your compound may depend on how well you understand what your efficacy model is actually testing and/or telling you. A mouse xenograft model derived from a human breast tumor cell or patient-derived tumor sample with a particular mutation driving its pathology may be a good model of that mutation-driven pathology, but is less likely to be a model of breast cancer. Likewise, mice with homozygous loss-of-function mutations in the genes encoding either leptin (ob/ob mice) or the leptin receptor (db/db mice) are both commonly used as obesity models in academic studies, but they model very different human patient populations. Depending on your drug's mechanism, it is critical to know which model to test your compound against.

A somewhat different situation exists for antiviral compounds under development to treat respiratory viruses, such as influenza or SARS-CoV-2. Animal "efficacy" studies with these compounds often test their ability to protect mice against death upon challenge with a lethal intranasal dose of the virus (or a mouse-adapted version of the virus). Although these studies can be used to demonstrate that the compound inhibits replication of the virus in the lung of an animal, as indicated by survival or reduction in virus titers in lung homogenates, they are not models of human disease and extreme care should be taken when trying to translate findings from these models to human studies. These studies are commonly performed as the only "efficacy" studies, however, since models more reflective of human disease (e.g., ferrets for influenza or ferrets or minks for SARS-CoV-2) are extremely expensive and challenging to run.

Hopefully these three examples together serve to underscore that it is essential to understand the details and limitations of each "disease-relevant" model as you decide how to evaluate the activity of your compound. Inability to run the "correct" study does not mean you should overinterpret data from the study you can conduct. Unfortunately, clinical development is unforgiving, and any errors made in interpreting the in vivo pharmacology studies will most likely come back to bite you.

> **Box 3.7: Important Topics Not Discussed in This Section**
> 1. **Need to understand the dynamic nature of your PD readouts**: In understanding PK/PD relationships it is essential to know whether your readout (PD marker) responds rapidly or slowly. This can hopefully be determined using cell-based assays, or can be estimated based on an understanding of the underlying biology.
> 2. **Need for PK information**: We typically assume that when we increase the administered dose in a dose-response study we increase drug exposure by the same relative amount, but this is often not the case. As such, we risk identifying a fully efficacious dose as being the dose beyond which we no longer increase drug delivery, not the drug exposure beyond which we can't improve efficacy. To guard against making this error, it is necessary to incorporate PK sampling into at least a subset of the in vivo pharmacology studies. Extensive PK sampling is not required since the goal is simply to estimate exposure; consider sampling at around the time of maximum plasma concentration (T_{max}) and then twice as plasma levels are expected to drop to roughly one quarter of maximum concentration (C_{max}). Regardless of where the intended site of action is, you should typically focus on plasma PK, because that is the tissue you will most likely sample in clinical studies.
> 3. **Biomarkers**: Biomarkers of pharmacodynamic activity or disease modification are extremely useful and becoming more common, especially in pharmacology studies. However, they are often hard to translate into the clinical setting, and there are extensive regulatory requirements if you plan to use them to make clinical decisions (Center for Drug Evaluation and Research (CDER) 2021). It's critical to understand the regulatory and practical requirements if you plan to use biomarkers in the clinic (see Sect. 2.11 for more on biomarkers).

3.3.4 Conclusions

In vivo pharmacology is a critical part of the drug development process, serving both as the final word in demonstrating that the selected molecule does what it is supposed to do and is worthy of consideration for clinical development, and hopefully providing some meaningful guidance to help the clinical group understand how to test the molecule in human trials. While there is general agreement on what studies should be done, there is less agreement on the order of these experiments and what data from those studies are most important. To be most successful, I suggest designing experiments to efficiently and effectively answer the questions of: (1) how much drug does it take to inhibit/activate the target, (2) what level of target modulation is required to get the desired downstream activity, (3) what fraction of the dosing interval is necessary to maintain this level of target modulation to achieve efficacy, and (4) how to select the patients most likely to benefit from the drug. When we have answered these questions, we have provided the most meaningful information to inform subsequent clinical development, and our predictions have likely proven correct in the clinic.

3.4   Pharmacokinetics and ADME

Werner Rubas and Emily Egeler

Initial screening efforts and secondary assays to identify compounds with desired efficacy and specificity for the intended target focus on issues of PD, which can be defined as "actions of a molecule (drug) on the body." For a drug to be successful, however, the active molecule must be able to reach the intended target at high enough concentrations and for a long enough time to exert its therapeutic effect. The body must also be able to remove the active molecule without significant buildup of toxic species, or the drug will fail in clinical trials. These considerations are evaluated in PK studies, summed up as "actions of the body on a molecule." PK studies measure the absorption, distribution, metabolism, and excretion of an administered molecule—often abbreviated as ADME.

3.4.1   Key ADME Parameters

For small molecules, ADME characteristics depend on i) properties of the molecule such as pKa/pI, molecular size/volume, and lipophilicity; and ii) subject characteristics such as age, body weight, sex, genetic variance (see Sect. 3.5 for more on the role of pharmacogenomics in ADME), organ impairment, and lifestyle factors such as diet, smoking, and co-medication. Excellent resources exist for detailed descriptions of the influence of each pharmacokinetic factor (see Sect. 3.4 Resources) (Rydzewski 2008). Important ADME characteristics are briefly discussed below and illustrated in Fig. 3.3 and Table 3.1.

In vitro ADME parameters:

- Plasma stability.
- Metabolic stability (intrinsic clearance-CL_{int}) obtained from liver specimens. CL_{int} is organ specific and can be assessed for other organs as well, for example, gut microsomes (also see below under in vivo).
- Plasma protein binding.

- Permeability including efflux.
- Substrate specificity for metabolic enzymes (phenotyping) and transporters.
- Enzyme (e.g., Cytochrome P450-CYP) and transporter inhibition/induction.
- Blood-to-plasma partitioning (concentration ratio between blood and plasma)— if >1, clearance based on plasma concentrations is high and misleading because of distribution into red blood cells.

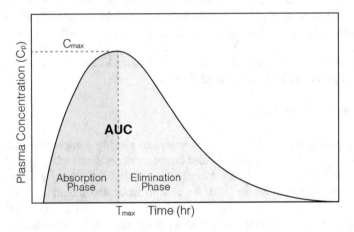

Fig. 3.3 Plasma drug concentration vs. time curve with pharmacokinetic metrics. Abbreviations: *AUC* Area Under the Plasma Drug Concentration vs. Time Curve, C_{max} Maximum Concentration, C_p Plasma Concentration, *t* Time, T_{max} Time of Maximum Plasma Concentration

Table 3.1 Useful pharmacokinetic (PK) equations. Abbreviations listed in table

Useful in vivo PK parameters with equations		
PK parameter	*Equation*	*Where…*
Total exposure	AUC	AUC = Area under the plasma drug concentration vs. time curve
Bioavailability	$\%F = \dfrac{AUC_{oral}}{AUC_{intravenous}} * 100$	F = Bioavailability AUC_{inf} = The area from time of dosing extrapolated to infinity
Blood/plasma clearance	$CL = \dfrac{Dose}{AUC_{inf}}$	CL = Clearance R = Renal
Renal/biliary clearance	$CL_R \, \& \, CL_{BL} = \dfrac{Ae_{0-t}}{AUC_{0-t}}$	BL = Biliary Ae = Total amount of drug eliminated in urine or bile per time interval t = time
Volume of distribution (initial distribution at time of dosing, no elimination)	$V_c = \dfrac{Dose}{C_0}$	V_d = Volume of distribution V_c = Central compartment C_0 = Concentration at time 0
Volume of distribution at steady state	$V_{ss} = CL * MRT$	V_{ss} = Volume of distribution at steady state
Half-life	$t_{1/2} = \dfrac{\ln 2 * V_d}{CL}$	MRT = Mean residence time

In vivo PK parameters—typically obtained from noncompartmental analysis (NCA):

- Model-independent PK parameters are derived from NCA and applied during drug discovery and development.
- Model-dependent (compartmental modeling) PK parameters are needed for simulation and/or prediction of efficacious exposures.
- Area under the plasma drug concentration vs. time curve (AUC)—the integral under a plot of plasma drug concentration versus time, where C_{max} is the peak concentration and T_{max} the time it takes to reach C_{max} (Fig. 3.3). The AUC reflects the "total exposure" from a single dose of drug and the dose normalized ratio of $AUC_{oral}/AUC_{intravenous}$ yields oral bioavailability (%F).
- Absorption (A) vs Bioavailability (F): The drug can be 100% absorbed, but the percentage of the administered dose that reaches the systemic circulation may be less due to metabolism and/or degradation during the absorption process from site of administration to systemic circulation.
- First pass effect: A phenomenon of drug metabolism in which an orally administered drug must pass through enterocytes (cells of the intestinal lining) and liver, undergoing biotransformation in the process. This reduces the effective concentration (and bioavailability) of the drug before it reaches systemic circulation and can be avoided by subcutaneous (sc) administration.
- Blood/Plasma Clearance (CL): A fraction of blood or plasma volume completely purified of drug per unit of time. Total CL is calculated as the ratio of Dose/AUC_{inf} and depends on the elimination rate constant (k_{el}) and V_d. Clearance at specific organs, such as liver, kidneys, skin, lungs, is dependent on organ blood flow; so disease states can alter drug clearance. Intrinsic clearance (CL_{int}) refers to the measured in vitro organ clearance.
- Renal/Biliary Clearance (CL_R & CL_{BL}): A fraction of blood or plasma volume completely purified of drug per unit of time in the kidney and/or liver. Clearance is calculated as the ratio of Ae_{0-t}/AUC_{0-t} with Ae_{0-t} total amount of drug eliminated during the collection interval and AUC_{0-t} the integral under the plasma curve during the interval.
- Volume of distribution (V_c, V_d, and V_{ss}): The apparent volume of distribution in the central compartment (V_c) with C_0 being the extrapolated concentration at time zero ($V_c = Dose/C_0$). Accuracy depends on adequate sampling time during the distribution phase. The apparent volume of distribution at steady state (V_{ss}) is a good estimate of the distribution between central compartment and tissue and is not dependent on rate of elimination ($V_{ss} = CL \cdot MRT$), where MRT is mean residence time (average time a molecule spends in the host). This is in contrast to V_d, which depends on the elimination rate (see also Half-life). For a drug retained exclusively in the vascular compartment, the volume of distribution is equal to the plasma volume (0.04 L/kg body weight). For a drug that is extensively bound in peripheral tissues, the V_d can greatly exceed the total body volume.
- Half-life ($t_{1/2}$): The time required for the drug concentration to fall by 50% of an earlier measurement. Terminal half-life is calculated from the clearance and volume of distribution [$t_{1/2} = ((\ln 2) \times V_d)/CL$].

3.4.2 Drug Metabolism and Drug–Drug Interactions

The simplest form of elimination is direct excretion of an unchanged drug molecule into the urine, bile, or occasionally tears, sweat, or air. More commonly, molecules undergo biotransformation, a process of metabolism that involves building or breaking chemical bonds within the molecule to improve the body's ability to excrete it. Biotransformation is grouped into Phase I and Phase II reactions. Phase I enzymes catalyze oxidations, reductions, and/or hydrolysis to introduce or unmask functional groups in the molecule. Phase II enzymes conjugate endogenous small polar molecules to the unmasked functional groups to inactivate the drug and improve its water solubility for elimination. A drug may be subject to Phase I metabolism, Phase II, or both. It is not uncommon that toxic products are formed during this process. Sometimes, knowledge of a drug's metabolism is exploited by chemists to devise a prodrug, a molecule whose metabolism creates the true therapeutically active compound, to improve ADME properties.

The CYP family of integral membrane (endoplasmic reticulum, cell, and mitochondria) enzymes is composed of many related isozymes and catalyzes the metabolism of about 80% of currently used drugs. These enzymes are responsible for a major portion of drug Phase I metabolism. CYP enzymes are expressed in all tissues and are most abundant in the liver, small intestine, and kidney. For instance, the CYP3A4 isoform is very abundant in the liver and intestinal epithelium and contributes to the biotransformation of almost one half of drugs, while CYP2D6 is one of the least abundant isozymes and is involved in the metabolism of a quarter of all drugs (Ekroos and Sjögren 2006).

Identifying which CYP isozymes are responsible for metabolism of the lead compound, called reaction phenotyping, is important for two reasons. First, many genetic polymorphisms have been identified for CYP isozymes. Polymorphisms result from inherited differences in enzyme expression or mutations that alter enzyme activity. These differences create variation in the rates of drug metabolism within a patient population. Dosing regimens may need to be adjusted to properly treat slow or ultra-fast metabolizers (see Sect. 3.5 on pharmacogenomics).

The second reason for reaction phenotyping is to understand the risk of drug–drug interactions. Co-administration of certain drugs or dietary components can impact the activity (inhibition) or abundance (induction) of the metabolizing enzyme(s), etc. These interactions must be carefully screened for, as they can either increase the clearance of concomitant medications, rendering them inactive, or decrease elimination and thereby result in toxic exposures.

3.4.3 AI and In Silico Experiments

Algorithms, a form of artificial intelligence (AI), have been developed to predict absorption, distribution, metabolism, excretion, and toxicity (ADMET) based on the structure of molecules (the inclusion of toxicity in ADME will be discussed below). Hundreds of molecules can be interrogated for their drug-like properties in a short period of time. This is a fast, efficient, and resource-saving approach to preselect molecules for chemical synthesis (Guan et al. 2019). The quality of prediction needs to be spot-checked with real experiments for chemical series. AI predicting drug–drug interactions, potential side effects, and the role of genetic variation on drug response can now be integrated into the early stages of drug discovery (see Sect. 2.6 on AI).

3.4.4   In Vitro Experiments

Initial studies of ADME characteristics are likely to be in vitro due to the high cost of animal studies. Although algorithms exist to extrapolate in vitro data to living systems, preliminary in vivo studies should be performed to confirm that in vitro data are indeed predictive. If the results are in concurrence, a strategy of in vitro screening with limited in vivo testing can be adopted. This approach allows more rapid and cost-effective identification of compound liabilities and better selection of a formulation before moving into animal models.

A number of different test systems are available to measure the in vitro or intrinsic clearance (CL_{int}), perform metabolite identification, and examine drug–drug interaction liabilities. These are listed in Table 3.2. CYP phenotyping is typically done using cDNA (complementary DNA) expression systems (SUPERSOMES™). In addition, a number of in vitro and culture models are used for drug metabolism (e.g., microsome assays) and to predict absorption via different routes of administration (Caco-2 and MDCK cell lines, mucosal tissues, and skin).

3.4.5   In Vivo Experiments

The goal of in vivo PK experiments is to calculate bioavailability, AUC, volume of distribution, and half-life while validating clearance and metabolite identity data collected from in vitro studies. For xenobiotics, FDA requires safety studies in at least two mammalian species, including one nonrodent species, typically the dog. A single species might be sufficient for biologics. Pharmacokinetic, toxicology, and

Table 3.2 In vitro test systems for intrinsic clearance

Test system	Specific models
cDNA CYP expression systems	Single enzyme system (Phase I and II, FMOs, MAO's, UGT, NAT's CESs)
Cell extracts	S9 fraction (Phase I and II)
	Microsomes (Phase I only)
Cell culture	Hepatocytes (fresh or cryopreserved)
	HepG2 cells transfected with CYP isozymes
Whole tissue	Liver slices

Abbreviations: *NAT* Arylamine N-Acetyltransferase, *CES* Carboxylesterase, *cDNA* Complementary DNA, *CYP* Cytochrome P450, *FMO* Flavin-Containing Monooxygenase, *MAO* Monoamine Oxidase, *UGT* UDP-glucuronosyltransferase pharmacodynamic studies are essential to set the starting dose for first in human (FIH) dosing. Typically, physiological-based pharmacokinetic approaches are used to predict the starting and efficacious doses for xenobiotics. For biologics, allometry in concert with PK/PD modeling provides the rationale for FIH dosing. Because upper dosing levels are usually set at the appearance of adverse side effects (or in the case of oncology drugs, severe adverse effects), in vivo pharmacokinetics studies go hand-in-hand with toxicology studies. For this reason, in vivo testing is often described as ADMET studies, in which the T refers to toxicity assays, including potential DNA damage and mutagenesis (Ames test), potential cardiac arrhythmia (cardiac hERG inhibition), and carcinogenicity liabilities; the latter three are tested in culture.

Rarely, a new molecule exhibits all the desired drug-like properties. In silico coupled with in vitro screening serve as a prelude to in vivo PK studies. Prior to initiating in vivo studies, predetermined criteria need to be met and confirmed in vitro. The single-dose PK (SDPK) design should aim to maximize the amount of information gained. Thus, collection of urine in addition to plasma is recommended. If the target is the central nervous system (CNS) or if CNS penetration is associated with undesired effects, brain exposures should also be determined. Dosing routes should include intravenous (IV) and the intended clinical route of administration, often oral (po). Plasma sampling should commence 2 min (IV) or 5 min (po) post dosing and continue for 24 h to enable creation of a profile as shown in Fig. 3.3.

Pay attention to the predicted ADME properties from in vitro studies. Any disconnect warrants an examination and decision whether optimization can proceed with predominantly in vitro experimentations or can only be successful with in vivo screening. Emerging lead molecules should also be examined for biliary excretion of parent compound and its metabolites. The compound may be secreted by the glycoprotein 1 (permeability glycoprotein, also known as multidrug resistance protein) into the intestinal lumen. The extent of its contribution to drug blood levels can be assessed in bile duct cannulated rats by collecting feces. A dose escalation study in the rodent toxicology species will establish the dose-limiting toxicity and the dose-limited absorption. If human efficacious levels cannot be attained, specialized formulations may help achieve the desired exposure.

3.4.6 The Bottom Line

Proper PK studies can help drug developers maximize their therapeutic window between minimum efficacious dose and maximum tolerated dose. PK studies inform how the lead compound is absorbed, distributed, metabolized, and excreted from the body. In vitro PK testing is used to identify initial metabolism rates and routes, in addition to identifying potential drug–drug interactions. In vivo PK testing is essential for establishing a pharmacokinetic/pharmacodynamic relationship and the maximum tolerated dose and therapeutic window in animals for a lead compound, which becomes the basis for planning safe and effective doses to move into human trials. Because different crystal forms, salts, and formulations of the same compound can have different ADME characteristics, it is very important to show favorable PK properties before scaling up good manufacturing practice (GMP) production for clinical trials to avoid costly reformulation delays.

> **Resources**
> - Rydzewski RM (2008) Chapter 9 – ADME and PK Properties. In: Real World Drug Discov. Chem. Guide Biotech Pharm. Res. Elsevier, pp 353–429.

3.5   Pharmacogenomics

Collen Masimirembwa

The term pharmacogenetics was coined in the 1960s to describe variability in drug response due to heredity. In the post-human genome era, the term pharmacogenomics appeared to capture the -omics approach with respect to biomarkers of drug response. Variations in genes that code for drug metabolizing enzymes and drug transporters affect the drug's PK, whereas variations in genes that code for drug targets influence the drug's PD. Ultimately, the effects on the PK or PD influence the individual's response to a drug with respect to safety and efficacy or both.

Many drug–gene interactions (DGI) have been characterized and some successfully used to optimize treatment outcome. This has led to the general belief that pharmacogenomics (PGx) is the low hanging fruit for genomics-based precision medicine (Dere and Suto 2009). The biopharmaceutical industry initially panicked at the perceived disruption of the traditional model of "one-drug fits all" model to one of individualized medicine that PGx was defining for the future of medicine. Cases like the antiretroviral drug efavirenz, which disproportionately caused an avoidable burden of neuropsychiatric adverse effects in people of African ancestry because of reduced metabolism, serve as a pointer that the pharmaceutical industry can do better with PGx knowledge (Fig. 3.4) (Masimirembwa and Dandara 2016).

> **Box 3.8: What Surprised an Academic?**
> Demonstrating a clear drug-gene-interaction predisposing people of African ancestry to efavirenz-induced adverse effects in 2008 was not enough! It took another 10 years for the clinical implementation of this knowledge. Unfortunately, this was too late for many patients whose lives were already affected by the drug's adverse effects. —*CM*

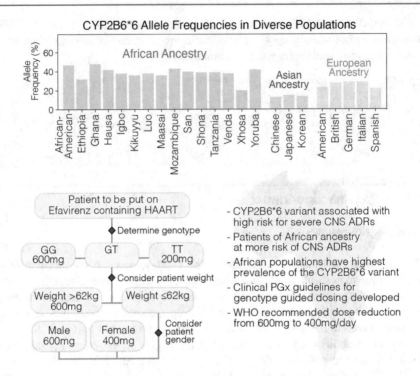

Fig. 3.4 Efavirenz risk for neuropsychiatric adverse effects. Figure adapted from (Masimirembwa and Dandara 2016). Abbreviations: *ADRs* Adverse Drug Reactions, *PGx* Pharmacogenomics, *CYP* Cytochrome P450, *CNS* Central Nervous System, *WHO* World Health Organization, *HAART* Highly Active Antiretroviral Therapy

3.5.1 Types of Genetic Variations and Technologies for Their Detection

The most common genetic variations in pharmacogenes are single nucleotide polymorphisms (SNPs). Although most polymorphisms are in noncoding regions or are synonymous (causing no amino acid changes), some of the nonsynonymous variations result in functionally important changes. Variations can also be due to insertions or deletions within the encoding genes, copy number variations that can result in loss of function, or gene amplification resulting in increased function. Finally, variations may also arise from epigenetic effects and post-translational modifications.

The best understood variations are in genes important for drug metabolism and pharmacokinetics. It is therefore not surprising that most DGI highlighted in FDA & EMA product labels and addressed by various clinical pharmacogenetics guidelines are for drug metabolizing enzymes. Whereas other genes can be important for specific drugs, variations in CYP2C9, CYP2C19, and CYP2D6 are among the most important for most drugs on the market. The Clinical Pharmacogenetics

Implementation Consortium (CPIC) provides a list of DGIs for which there is very strong clinical evidence (Level 1A) to support their application in the use of some drugs to reduce risk for adverse effects and/or increase treatment efficacy (see Sect. 3.5 Resources).

Genetic variations in drug targets should also be considered. Here, in addition to germline variations, somatic cell variations in genes coding for drug target pathway proteins have gained prominence. Genetic variation in key signaling pathways in cancer biology such as HER2, KRAS, EGFR, ALK, and BRAF can impact the efficacy of many monoclonal antibodies such as trastuzumab, panitumumab, crizotinib, and trametinib. Use of these drugs strongly depends on the availability of genetic tests for the biomarkers since the mutation profiles can change during the course of disease progression and treatment, necessitating change of therapeutic agent combinations to ones more likely to work in the evolving tumor (Shek et al. 2019).

Initially, PCR-restriction fragment length polymorphism (RFLP) methods were used for the detection of PGx biomarkers, but have been replaced by reverse transcription (RT)-PCR used by most molecular diagnostic laboratories, especially in multiplex formats using open array and chips. This in turn will be replaced by next generation sequencing (NGS) when it becomes less expensive. The choice of genotyping platform depends on the application.

The use of SNP panels is very common since it is fast, easy to implement and interpret, and inexpensive. Unfortunately, it is limited in coverage for other variants—especially from different ethnic groups. Sequencing gene SNP panels detects rare variants and is still relatively inexpensive, but is also limited to the targeted genes on the sequence panel. Genome-wide genotyping, a more expensive approach, is useful for population studies (Genome Wide Association studies (GWAS)), but have the challenge of requiring large sample sizes to detect statistically significant drug–gene interaction associations. NGS enables the discovery of novel variants, including variants of unknown clinical significance (VUS) that are not clinically actionable. To apply the findings to patient care, PGx requires analytical validation of the methods, performing tests in accredited laboratories, and approval of tests by the relevant regulatory authorities (Arbitrio et al. 2021).

3.5.2   Pharmacogenomics in Drug Discovery and Deployment

In the drug discovery, development, and deployment value chain depicted in Fig. 3.5, pharmacogenomics now plays an important role. This is evidenced by the numerous genomics research programs in biopharmaceutical and biotechnology companies and by the long existing Industry Pharmacogenomics Working Group (I-PWG) (Bienfait et al. 2021). The role of genetic variations in drug targets, which has been demonstrated often in the development of anti-cancer drugs, and how the study of these variations should be applied in the discovery phase or in patient selection for clinical studies, is not discussed here.

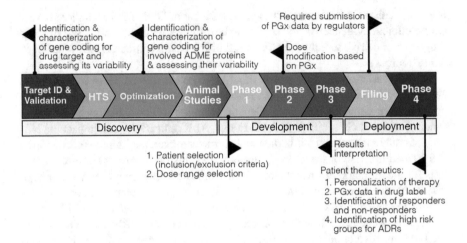

Fig. 3.5 The role of pharmacogenetics in the drug discovery, development, and deployment value chain. Abbreviations: *PGx* Pharmacogenomics, *ADME* Absorption, Distribution, Metabolism, and Excretion, *ADRs* Adverse Drug Reactions, *HTS* High-Throughput Screening, *ID* Identification

3.5.2.1 Late Discovery Phase

During lead optimization and drug candidate selection, standard in vitro assays to identify drug metabolizing enzymes and drug transporters that are involved in the disposition of new chemical entities can be used and include cellular, subcellular, and recombinant proteins (see Sect. 3.4 for more on ADME/PK). Identifying metabolizing enzymes with genetic variability, such as CYP2B6, CYP2C9, CYP2C19, and CYP2D6 alerts the drug discovery team to select other analogues and/or to increase medicinal chemistry efforts to reduce the role of the polymorphic enzymes. The team may also proceed with the compound, but further phase 1 clinical pharmacogenetic studies will be needed to quantify the pharmacokinetic impact of the genetic variability on the specific compound.

3.5.2.2 Development Phase

If the ADME of a new chemical entity is affected by polymorphic genes, this should be considered when designing the phase 1 study to allow careful evaluation of the effects of the genetic variations. For example, if the compound mainly depends on CYP2D6 for its metabolism, participants with a CYP2D6 poor metabolizer (PM) genotype may be excluded or given a relatively low dose of the test compound. Such stratification prevents or reduces the occurrence of severe adverse drug reactions (ADRs) and can contribute to better clinical trial outcomes, thus reducing attrition rates.

If PGx studies are included in phase 1 and phase 2 studies, much failure related to both lack of efficacy and/or toxicity can be avoided. PGx documentation in phase 3 should be included along with other confounding factors, such as different co-morbidities, co-treatments, different age groups, and importantly different population ancestry groups. It has now become standard to biobank DNA samples from clinical trials, as this affords industry an opportunity to investigate potential genetic factors to explain clinical trial observations beyond those that could have been anticipated from phase 1 and 2 PGx studies. Regulatory agencies such as FDA and EMA have provided recommendations regarding how to submit pharmacogenomic data in the New Drug Application (NDA) dossier (Sect. 3.5 Resources). This information helps the regulators with product label information. Over 150 products on the market now have some PGx information and/or clinical guidelines (Sect. 3.5 Resources).

Although regulatory authorities have provided PGx product label information, extensive clinical studies and evidence evaluation is needed to inform clinical guidelines on how to implement PGx in the clinical setting. Many consortia of experts have been developing such guidelines. Notable ones include CPIC, the Royal Dutch Association for the Advancement of Pharmacy—Pharmacogenetics Working Group (DPWG), the Canadian Pharmacogenomics Network for Drug Safety (CPNDS), and the French National Network of Pharmacogenetics (RNPGx). The PharmGKB database then keeps an accessible summary of these guidelines on its website (Sect. 3.5 Resources).

There are now clinical guidelines for over 148 drugs on the market. Guidelines from different consortia are not always in agreement, and there are ongoing efforts to harmonize them to ensure their generalizability. Unfortunately, product labels and clinical guidelines are mainly based on DGI data obtained in populations of European ancestry, with increasing data on Asian populations, but very little data on people of African ancestry, since very little clinical PGx research has been done in Africa (Radouani et al. 2020). Even the most commercially available PGx tests, such as for CYP2C9, do not test for the African-specific variants, such as CYP2C9*5, *6, *8, *11; therefore, results from those tests are not predictive for dosing of common drugs such as warfarin or siponimod in people of African ancestry. The same is true for CYP2D6*17 and *29, for which African-specific variants are not included, resulting in poor dose predictions for African populations. To address this deficit in data and knowledge, a number of programs have been initiated in Africa, such as iPROTECTA (implementing pharmacogenetic testing for effective treatment and care in Africa) that use PGx tests such as GenoPharm that are inclusive of variants unique to African populations (Mbavha et al. 2022). These studies are inspired by the highly successful PREPARE program in Europe (van der Wouden et al. 2020) and IGNITE in America (Weitzel et al. 2016).

These global programs are also working to address barriers in clinical implementation of PGx such as knowledge, attitudes, and practices (KAP), ethical legal and societal issues (ESI), laboratory facilities, electronic patient record systems, and clinical decision support systems (CDSS) for point-of-care application of PGx (Roosan et al. 2021). Whether the deployment of clinical PGx uses a preemptive or a reactive approach, it should result in safer and more effective use of medicines and fewer failures in drug development (Fig. 3.6).

Integrating PGx in national pharmacovigilance programs can be an invaluable tool to improve safety and efficacy for medicines with PGx issues as they are deployed across the world. For example, analyzing ADR data deposited in the

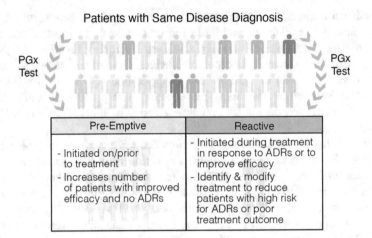

Fig. 3.6 The anticipated impact of clinical pharmacogenetics on the use of some medicines if adopted as the standard of care. Abbreviations: *ADRs* Adverse Drug Reactions, *PGx* Pharmacogenomics

World Health Organization (WHO) database, Vigibase, shows that for some drugs, different populations report different risk for ADRs (Ampadu et al. 2016). Such approaches could help regulatory agencies identify medicines that require bridging pharmacokinetic and pharmacogenetic studies in different population groups.

The Bottom Line
- Pharmacogenomics (PGx) is the field that studies variations in genes that code for drug metabolizing enzymes and drug transporters.
- PGx plays an important role in drug development. If ignored, it can put good discovery programs at risk for failure in later stages of development. It is therefore highly recommended to integrate PGx testing early in drug development.
- When drugs are approved, barriers in clinical implementation of PGx include lack of knowledge, incorrect attribution to cultural differences rather than genetics, and ethical, legal, and societal issues. Other barriers for applying PGx information for better patient care include lack of laboratory facilities, challenges with electronic patient record systems, and lack of clinical decision support at the point of care.
- Unfortunately, drug labels and clinical guidelines are mainly based on drug–gene interaction data obtained in populations of European ancestry, with increasing data on Asian populations, but very little data on people of African ancestry.

Resources
- CPIC Gene-Drug Interactions: https://cpicpgx.org/genes-drugs/.
- FDA Guidance for industry, pharmacogenomic data submission document: https://www.fda.gov/media/122944/download.
- FDA Table of Pharmacogenomic Biomarkers in Drug Labeling: https://www.fda.gov/drugs/science-and-research-drugs/table-pharmacogenomic-biomarkers-drug-labeling.
- PharmGKB Clinical Guideline Annotations: https://www.pharmgkb.org/guidelineAnnotations.

3.6   Route of Administration and Drug Formulation

Terrence F. Blaschke

The route of administration and the formulation of a drug are dependent on the chemical composition of the drug and the desired onset and duration of action. The route of administration of a drug can be broadly separated into three categories: (1) enteral, (2) parenteral, and (3) topical (Table 3.3). In each of those categories, there are a number of subcategories.

Each of these routes of administration requires a different type of formulation. Many companies are developing drug delivery technologies involving oral, nasal, inhalation, transdermal, and parenteral delivery platforms.

Box 3.9: What Frustrated an Academic?
Not all drugs reach their target when delivered in a simple formulation. Proper formulation and route of delivery is also critical when using new pharmacological agents for basic research, whether in culture or in vivo. It is important to include studies on drug stability and distribution for each formulation of a new pharmacological agent. —*DM-R*

3.6.1   Oral Route

The most common, desirable, and usually the least expensive route of administration is the oral route, especially if the drug is intended for multiple doses or chronic administration. However, for many drugs, the oral route may not be feasible or practical, as the drug may have poor oral bioavailability and not reach adequate concentrations in the systemic circulation. For the oral route, there are many different forms (tablet, capsule, liquid, suspension, etc.) chosen based on the convenience of the formulation and the bioavailability of the drug.

Another important characteristic of an oral formulation is its rate of absorption. In some settings, rapid absorption is desirable to achieve a rapid onset of action

Table 3.3 Three major categories for route of administration

Route of administration categories		
Enteral	*Parenteral*	*Topical*
Absorption through the GI tract	*Absorption outside the GI tract*	*Absorption at the site of application, such as skin or mucous membranes*
Oral	Intravenous	Transdermal (intended for systemic effects)
Buccal or sublingual	– Slow bolus	
Rectal	– Slow infusion, then stop	Epidermal/dermal (intended for local effects at the site of administration)
	– Continuous infusion (long term)	Vaginal (usually intended for local effects)
	Subcutaneous	
	– Bolus	Intranasal (intended for local or systemic effects)
	– Continuous infusion (long term, e.g., insulin)	Pulmonary inhalation (intended for local or systemic effects)
	– Depot	
	Intramuscular	
	– Bolus	
	– Depot	
	Intraocular	
	– Bolus	
	– Depot	

Abbreviations: *GI* Gastrointestinal

(e.g., drugs given for pain or for sleep). Tablets may be formulated as "quick dissolve" versions. In other settings, rapid absorption is problematic, as the high peak concentrations associated with rapid absorption may result in unwanted side effects, sometimes serious or life-threatening.

A number of special formulations used for oral administration are intended to prolong the duration of action and/or avoid high peak concentrations. These are often called "slow release" (SR) or "extended release" (XR) formulations to distinguish them from immediate release formulations. Such formulations may allow a drug to be administered at longer dosing intervals that improve patient adherence to the medication (e.g., once instead of twice daily). Other special oral formulations include enteric-coated formulations that protect the drug from the acidic environment of the stomach and dissolve in the intestines, or fixed-dose combinations containing two or more active pharmaceutical ingredients (APIs) that are used for conditions benefiting from combined drug therapy (e.g., hypertension, diabetes mellitus, and HIV).

3.6.2   Parenteral Route (Injectables)

For drugs that cannot reach the systemic circulation after enteral or transdermal administration, or for drugs for which a very rapid onset of action is needed, parenteral dosage forms are required. Parenteral routes also avoid the first-pass metabolism in the liver experienced by orally administered drugs. For direct IV administration, the drug must be solubilized in a liquid suitable for direct injection into a vein. Speed of injection (bolus, slow infusion, or constant infusion) is dependent on the indication. For anesthetics and sedative/hypnotics used in procedures and for some cardiac arrhythmias, slow bolus injections are often used. However, for many other agents that are not orally available (e.g., many anticancer agents and the rapidly increasing number of biologics on the market), a slow infusion is preferable to avoid toxicity associated with high peak concentrations and rapid distribution into tissues where unwanted effects can occur (e.g., the central nervous system, the heart, or other vital organs). The advent of reliable, miniaturized infusion pumps is increasing interest in research evaluating whether the therapeutic index could be improved by longer-term infusions. The subcutaneous infusion of insulin is an example of this approach in treating diabetes.

3.6.3 Epidermal or Transdermal Route

Epidermal or transdermal formulations are generally patches or gels. If systemic absorption is the goal of transdermal delivery, many characteristics of the drug may limit this route. Drugs must be of high potency, be able to penetrate the epidermis, and benefit from a constant concentration in the blood. Alternatively, transdermal or epidermal routes may be selected to deliver a high local concentration of drug and avoid systemic exposure. There is increasing interest in this route of administration. Patches are easy to use (improving patient adherence), provide continuous dosing of a steady drug concentration, and avoid first-pass metabolism by the liver. A number of companies are developing new technologies to improve transdermal absorption. A few examples of successful transdermal systemic delivery systems include the opiate pain reliever fentanyl, contraceptive patches, and clonidine for hypertension. Examples of successful drugs used for local effects include topical steroids, antibiotics, and local anesthetics.

> **Box 3.10: What Surprised an Academic?**
> A formulation consultant suggested that we formulate our intracoronary drug at pH 3, as the drug was more stable in acidic conditions. Supporting his arguments, he cited a few drugs on the market that use this route and are formulated at a low pH. Luckily, our clinical director knew that the drugs mentioned produced phlebitis (local infection of the vein) and helped me, the basic researcher, to push back on that formulation recommendation. Consultants are not always right and, if something does not seem right, we should do our own diligence. —DM-R

3.6.4 Biologics Require New Delivery and Formulation Methods

A recent market analysis valued the current biologics market at $343 billion, and this is expected to increase to nearly $570 billion by 2027 (Biologics Market 2022).

This rapid increase in the number of biologics on the market and in the pipeline has provided incentive for the development of new technologies to improve their delivery and efficacy/safety. As a result, there is an emerging market to devise parenteral formulations to produce so-called "bio-betters" that require less frequent administration and have an improved therapeutic index. A recent survey found that more than 20 independent drug delivery companies are researching controlled release depot injection formulations, and many major pharmaceutical companies have internal programs. These formulation technologies to deliver biologics include microspheres, liposomes, microparticles, gels, and liquid depots (see examples listed in Sect. 3.6 Resources). Currently, there are 13 depot products on the market, and the market size for such products is estimated to be >$2 billion.

Box 3.11: What Surprised an Academic?
When I first started doing clinical research at Stanford, I realized that the most important and common cause of a poor outcome was not the drug or its formulation, but poor patient adherence. This fact impacted the rest of my career. Keep in mind the statement of former US Surgeon General C. Everett Koop in 1985: "Drugs don't work in patients who don't take them". —*TB*

The Bottom Line
Identifying an optimal drug formulation involves a process of trial and error. Success requires an in-depth understanding of the drug's PK/ADME, the desired patient exposure profile, and the clinical setting. Compromise may be required to balance the chemical properties of the API with the desired pharmacodynamic effect.

Resources
Reviews:
- Liechty WB, Kryscio DR, Slaughter BV and Peppas NA (2010) Polymers for Drug Delivery Systems. Annual Review of Chemical and Biomolecular Engineering. 1:149–173. (Broad and comprehensive review of this topic containing 149 references and other related resources.)
- Wang AZ, Langer R, Farokhzad OS (2012) Nanoparticle Delivery of Cancer Drugs. Annual Review of Medicine 63:185.
- Timko BP, Whitehead K, Gao W, Kohane DS, Farokhzad OC, Anderson D, Langer R (2011) Advances in Drug Delivery. Annual Review of Materials Research 41:1. (This review discusses critical aspects in the area of drug delivery. Specifically, it focuses on delivery of siRNA, remote-controlled delivery, noninvasive delivery, and nanotechnology in drug delivery.)

Books:
- Derendorf M, Schmidt S (2020) Clinical Pharmacokinetics and Pharmacodynamics: Concepts and Applications, 5th Edition. (Study questions are found at the end of each chapter with answers at the end of the book. The text is accompanied by web-based material, including computer-based simulations of many concepts.)
- Principles of Clinical Pharmacology—4th edition by Arthur J. Atkinson. Publisher: Academic Press, Inc. Published: 2022.

Websites:
- National Institutes of Health, Office of Clinical Research, Clinical Research Training, Principles of Clinical Pharmacology. A free, online course, taught by faculty members from the National Institutes of Health (NIH) and guest faculty from FDA, the pharmaceutical industry, and several academic institutions across the US: https://ocr.od.nih.gov/courses/principles-clinical-pharmacology.html.
- American College of Clinical Pharmacology, Continuing Education. Online educational platform teaching various disciplines related to drug-human interactions (membership fee): https://www.accp1.org/Members/ACCP1/4Continuing_Education/Continuing_Education.aspx?hkey=30506969-1528-454b-a09c-54b8063043dd.
- University of Maryland, Center for Translational Medicine. A free, web-based learning resource covering key concepts in pharmacometrics: https://ctm.umaryland.edu/#/ms-pharma.

3.7   Preclinical Safety Studies

Michael Taylor and Kevin Grimes

"Primum non nocere" translates from Latin to "First, do no harm." This fundamental ethical principle in the practice of medicine is equally applicable when exposing individuals to investigational drugs. Virtually all substances can be toxic to human beings if the dose is high enough. Even drinking excessive quantities of water or breathing 100% oxygen for prolonged periods can result in severe organ damage or death. Therefore, when administering a novel compound to human subjects, we have both an ethical and legal duty to ensure that the risk has been minimized as much as possible.

Safety is difficult to prove without extensive human exposure. Lack of safety, on the other hand, can be proven. We perform preclinical safety studies to better characterize the likely effects and the risk/benefit ratio of administering a novel compound to humans. Whereas experiments using cell lines and animal models will not mirror with certainty what will happen in human subjects, the results can be extremely helpful in predicting dose-limiting side effects and appropriate dose ranges.

FDA and the International Committee on Harmonization (ICH) have developed guidance documents that outline a series of in vitro and in vivo experiments that should be conducted prior to each phase of clinical development for a new molecular entity (NME). These studies help predict the drug's on-target and off-target toxicities, reversibility of these toxicities, limits on the dose and duration of treatment, early predictors or signals of impending serious toxicity, and the safety margin between doses where efficacy and dose-limiting toxicity occur. Additional studies are performed to further characterize the drug's pharmacologic effects on major organ systems, pharmacokinetics, metabolism, and likely interactions with food or other drugs. Preclinical safety studies that will be submitted to regulatory agencies to support subsequent clinical testing must be performed according to Good Laboratory Practice (GLP). GLP studies require extensive documentation of each study procedure and are quite costly.

It is important to recognize that an acceptable safety profile will be determined according to the potential risk versus the potential benefit related to drug exposure. The number of individuals exposed to the toxic effects of a drug may also be factored into the risk/benefit calculation. For example, there will be a much higher tolerance for adverse effects from a new drug for an incurable, life-threatening disease than for a novel antihistamine. Perhaps the highest safety burden is placed on new prophylactic vaccinations that are administered to broad populations, including

---

Box 3.12: What Surprised an Academic?

The drug tested in GLP toxicity studies should not be too pure. If the clinical lot has higher levels of impurities than the toxicology lot, which can occur from manufacturing scale-up, further GLP toxicology studies will be required to characterize the potential toxic effects of the new or increased impurities. This can significantly impact development timelines and budgets. So it took me some time to understand, when told by the VP of Drug Development, that my pride in purifying our non-GLP material to 99.5% purity before using it in pig efficacy studies was misguided and potentially a very costly mistake. —DM-R

---

children, to prevent a potential illness rather than to treat an existing one (see Sect. 2.10 for an expanded discussion). The WHO guidance on nonclinical evaluation of vaccines provides recommendations for the development of prophylactic vaccines (World Health Organization 2005).

Although there is always opportunity for discussion and negotiation, FDA (and other regulatory agencies) typically requires a specific battery of nonclinical safety studies to be completed before advancing to phase 1 human studies. In general, the duration of drug exposure in animal studies should equal or exceed that of subsequent clinical studies. Therefore, additional general animal toxicology studies of longer durations are often performed to support increasing duration of clinical dosing prior to phase 2 and phase 3 studies. Specific studies of relatively long duration assessing reproductive toxicity and carcinogenicity may be required before exposing large numbers of patients to study drugs in phase 3.

The guidance documents include discussions of various types of studies to assess specific toxicities including safety pharmacology of the cardiovascular, pulmonary, and neurologic systems; genotoxicity; reproductive toxicity; and carcinogenicity. In addition, they outline preclinical safety requirements for specific disease indications (e.g., oncology).

The requirements for biotechnology-derived products and vaccines differ from the requirements for traditional drugs. The safety evaluation of biologics (e.g., monoclonal antibodies) can be limited by a lack of therapeutic or target homology between the animal model species and humans. Likewise, the duration of testing may be limited by the development of neutralizing antibodies to the therapeutic. Therapeutic vaccines are typically comprised of an endogenous (e.g., cancer cell) antigen and are developed to treat an existing serious disease in a limited number of patients. Conversely, prophylactic vaccines have foreign (e.g., bacterial, viral) antigens and are administered to broad populations of relatively healthy individuals. Therefore, the accepted risk/benefit ratios and the specifics of study design will vary significantly. The animal species used for vaccine safety testing should be capable of developing an immune response to the vaccine antigen. The extent and duration of the in vivo immune response may help predict the frequency of human dosing that must also be evaluated in preclinical safety studies.

In addition to identifying possible toxicities, nonclinical safety studies are important for identifying potential biomarkers to monitor untoward effects, establishing the first dose to be administered to humans, and establishing the upper limits of dosing (exposure) in humans. This latter purpose is particularly important when severe or nonmonitorable toxicities are encountered.

Guidance regarding the development of previously approved drugs for new indications, by comparison, is limited. There is a specific guidance that indicates the kind of animal studies required for reformulated old drugs (also termed repurposing or repositioning). FDA also expects that an old drug being developed for a new indication meets current regulatory standards.

Before conducting animal studies, it is important to define how the drug will be given to patients: formulation, route of administration, and frequency of dosing. Generally speaking, animal testing of drugs intended for nonoral routes should make use of the same formulation and route of dosing that will be used clinically. Both the excipients (inactive ingredients of the final formulation) and API and associated impurities need to be considered and evaluated. It is important to appreciate that excipients are scrutinized during the approval process similarly to the drug under development.

When determining which excipients to include in the final formulation, the FDA inactive ingredients listing can be useful. A novel excipient or novel use outside the limits of its current use (e.g., route, dose) will normally require additional evaluation. The use of some excipients is limited by toxicity (e.g., dimethylacetamide, cyclodextrin), and therefore, it is necessary to carefully consider the excipient dose and the patient population for which the product is intended. A good strategy for excipient evaluation is to use the clinical formulation without API as the vehicle formulation (control group) in animal studies. It is also advisable to include an additional negative control group to confirm lack of effects by the excipient.

The selection of the API lot for animal testing is also important. The tested material should be representative of the material intended for clinical use, such that the impurity profile is both qualitatively and quantitatively similar to the clinical material. Several guidances discuss the acceptable limits of API impurities and the necessary steps for impurity qualification when such limits are surpassed. A good practice, particularly for the IND-enabling studies, is to use the same lot of API for nonclinical safety studies that is to be used in the clinic.

Appropriate dose selection is important to the conduct of successful animal toxicology studies. In part, success should be considered based on efficient use of animals. Although the use of two species of animal models is central to drug development and evaluation, there is an ever-increasing awareness and responsibility to follow humane practices and to thoroughly justify the need for animal use and numbers.

The fundamental premise of dose selection for animal studies is that the animal doses and exposures (C_{max}, AUC) should exceed those proposed for humans. Ideally, the high dose for animal studies is best selected by clear evidence of toxicity, such as decreased body weight, changes in clinical condition, or abnormalities in clinical or anatomic pathology parameters. The low dose should be a small multiple (2–3×) of the projected clinical dose (exposure) and the mid-dose should be set between the high and low doses. It is important to separate doses such that the exposures between groups do not overlap. For many orally delivered small molecules or parenterally delivered macromolecules, doses can be adequately spread using half log or log intervals. Since there is less pharmacokinetic variability for intravenous administration, the dose intervals can be smaller.

Because both dose and time influence toxicity, it is difficult to predict doses that will be tolerated for chronic administration. Therefore, it is best to plan studies of increasing duration sequentially. Selection of doses for the first studies can be challenging, and you should draw on all available information. Whereas rodents are usually the species chosen for early efficacy studies, there is typically no information available for dosing in the nonrodent model. If no or limited data are available, short duration non-GLP pilot studies (1–3 days) evaluating toxicity and plasma drug concentrations in a minimal number of animals should be performed to help select the appropriate dose range. For compounds with limited evidence of toxicity, the high dose can be set based upon consideration of the animal exposure relative to humans and practical limits such as dose volume or API solubility.

This discussion introduces the types and extent of preclinical safety studies required to support drug development. You should also consult the previous sections on formulation and drug metabolism, as these are also important considerations for successful safety evaluation.

Resources
- Specific FDA guidance on Nonclinical Safety Studies: http://www.fda. gov/downloads/Drugs/GuidanceComplianceRegulatoryInformation/ Guidances/UCM073246.pdf.

The Bottom Line

When administering a novel compound to human subjects, we have both an ethical and legal duty to ensure that the risk has been minimized as much as possible. Preclinical safety studies help to minimize risk to human subjects by identifying potential toxicities, appropriate dosing ranges, and early signals of toxicity.

Key Terms and Abbreviations

Key Terms

Absorption, Distribution, Metabolism, and Excretion (ADME): PK parameters that measure how the body processes drugs.

Active Pharmaceutical Ingredient (API): The ingredient that produces the intended effects, also called drug substance.

Adverse Drug Reactions (ADRs): Unwanted or dangerous side effects caused by the administration of a drug.

Alzet® Pumps: Miniature osmotic infusion pumps for continuous dosing of a drug to a laboratory animal.

Ames Test: An assay performed in bacteria that tests whether a compound causes mutations in the DNA of the test organism.

Area Under the Plasma Drug Concentration vs. Time Curve (AUC): Plasma concentration of a drug integrated over time after dosing.

Artificial Intelligence (AI): The use of a machine/computer to mimic human intelligence by executing tasks normally performed by humans.

Arylamine N-Acetyltransferase (NAT): A polymorphic enzyme that catalyzes the transfer of an acetyl group (from acetyl-CoA) to a xenobiotic acceptor containing aromatic amines. Acetylation of drugs enhances excretion.

Bio-Betters: New formulations of biologic therapeutics to improve dosing schedule or route of administration.

Bioavailability: Based on the route of administration, the fraction (or percent) of the dose of chemically unchanged drug reaching systemic circulation.

Bolus: A single dose of an injected drug given over a short period of time.

Buccal: In the mouth.

Carboxylesterase (CES): A class of enzymes that catalyze the hydrolysis of esters, amides, thioesters, and carbamates. Their most common drug substrates are ester prodrugs, though they can also convert xenobiotics into inactive metabolites.

Center for Drug Evaluation and Research (CDER): Center within the US Food and Drug Administration (FDA) responsible for the regulation of

over-the-counter and prescription drugs (including some biological therapeutics and generic drugs).

Compartmental Modeling: Requires log transformation of the plasma concentration data. Each linear elimination phase is ascribed to a fictive compartment (e.g., one phase = 1-Comp model). A mathematical model is
developed to describe the log-transformed plasma concentrations.
Assuming linear PK, this model can be used to simulate different doses
and/or schedules.

Cytochrome P450 (CYP): A class of heme-containing enzymes (monooxy-
genases) involved in drug metabolism. Genetic polymorphisms exist in
subpopulations with altered abilities to metabolize xenobiotics (e.g., poor
metabolizers).

Depot: A drug formulation injected into the body that serves as a reservoir
allowing drug release over an extended period of time, measured in days
to months.

Drug–Gene Interactions (DGI): The relationship between a drug and the
genetic makeup of an individual that may affect the patient's response to
the drug.

Endpoints: Measurements (e.g., weight or tumor size) or observations (e.g.,
motor control or healthiness) used in a study to evaluate the effectiveness
or safety of a treatment.

Enteral: Routes for drug absorption through the gastrointestinal tract.

European Medicines Agency (EMA): An agency of the European Union
responsible for the evaluation, supervision, and safety of medicinal products. EMA and US FDA have similar missions to ensure that medical products are both safe and efficacious before they are approved for marketing.

Excipients: Inactive materials (e.g., fillers, binders, coatings) included in the
drug product formulation.

Extended-Release (XR): An oral formulation that releases its API over 1 day.

First in Human (FIH): A clinical trial in which a treatment is tested in
humans for the first time.

Flavin-Containing Monooxygenase (FMO): A class of enzymes that adds
molecular oxygen to lipophilic xenobiotics, enhancing solubility and rapid
elimination of the molecule.

Good Laboratory Practice (GLP): Extensive documentation of each procedural step to ensure high quality, reproducible studies.

Good Manufacturing Practice (GMP): Exacting procedures and documentation of quality assurance carried out at a certified drug manufacturing
facility (sometimes referred to as "cGMP" for "current" practice).

Human Equivalent Dose (HED): A calculated dose predicted to achieve the
same exposure in humans as in a relevant animal model. This is a species-
dependent ratio based on body surface area that generalizes how a given
dose in that species will convert to a corresponding human dose. These

ratios can be found in the FDA guidance document on First in Human Testing.

Human Ether-á-Go-Go-Related Gene (hERG) Channel: A potassium ion channel that is important for normal electrical activity of the heart. Inhibition of this channel can lead to sometimes fatal cardiac arrhythmias.

International Committee on Harmonization (ICH): Joint effort of European, Japanese, and US regulatory authorities and pharmaceutical industries to provide uniform standards and guidance for drug development.

Intranasal: In the nose.

Intraperitoneal (ip): Within the peritoneum (abdominal cavity).

Intravenous (IV): Within a vein.

Investigational New Drug Application (IND): Document filed with FDA prior to initiating research on human subjects using any drug that has not been previously approved for the proposed clinical indication, dosing regimen, or patient population.

Maximum Concentration (C_{max}): The maximum concentration a drug achieves in a given body fluid or tissue (most often plasma) after the drug has been administered.

Monoamine Oxidase (MAO): A family of enzymes that catalyze the oxidation of monoamines, implicated in brain metabolite clearance (serotonin, dopamine, norepinephrine).

New Drug Application (NDA): Application to obtain FDA approval for the sale and marketing of a new drug in the US.

New Molecular Entity (NME): A novel drug that has not been previously approved for human use.

Noncompartmental Analysis (NCA): A type of PK analysis commonly used in drug development. This method is often used for assessing clearance, half-lives, and distribution.

Oral (po): By mouth.

Orphan Indication: An FDA designation of a disease or condition that affects less than 200,000 people per year in the US or for a treatment that is not expected to recoup its research and development (R&D) costs due to pricing constraints.

P-Value: A statistical measure of the probability of obtaining a result at least as extreme as the one observed. Although there is no experimental basis for doing so, it is routinely accepted that if the p-value is less than the significance level (usually 0.05 or 0.01), one rejects the null hypothesis that there is no treatment effect. The "requirement" for studies to achieve $p < 0.05$ is a major contributor to research waste (Begley and Snapinn 2021).

Parenteral: Routes for drug absorption outside the gastrointestinal tract.

Pharmacodynamics (PD): Measurements of drug action in the body (e.g., target inhibition/activation, cell killing, change in blood pressure).

Pharmacogenomics (PGx): The field that studies variations in genes that code for drug metabolizing enzymes and drug transporters.

Pharmacokinetics (PK): Measurements of what the body does to a drug (absorption, distribution, metabolism, and excretion).

Pharmacokinetics/Pharmacodynamics (PK/PD): Interplay of pharmacokinetics and pharmacodynamics that ultimately determines the biologic impact of a drug (i.e., whether a drug will be safe and efficacious for the disease indication as dosed).

Preclinical Animal Studies: Animal studies performed to validate a disease target and test the performance of a molecule prior to moving into human testing.

Single Dose Pharmacokinetics (SDPK): A dosing regimen in which a single dose of a compound is administered and PK is measured. SDPK can provide a relatively robust understanding of the PK/PD relationship of a compound's effect on its target.

Single Nucleotide Polymorphism (SNP): A single DNA nucleotide base change in the genome that contributes to genetic variation.

Slow-Release (SR): An oral formulation that releases its API over a period of hours. SR formulated drugs are often dosed every 12 h.

Structure–Activity Relationship (SAR): The relationship between the chemical structure of a molecule and its biological activity. During lead optimization, iterative chemical modifications are made to a molecule and the impact on various biological parameters is assessed.

Subcutaneous (sc): Beneath the skin.

Sublingual: Under the tongue.

Therapeutic Index: The ratio of the toxic dose to the effective dose; a larger therapeutic index suggests a larger safety window.

Time of Maximum Plasma Concentration (T_{max}): The time at which the drug reaches its maximum concentration after administration.

Topical: Routes for drug absorption at the site of application, usually the skin or mucosal membranes.

UDP-Glucuronosyltransferase (UGT): A class of enzymes that conjugate glucuronic acid to lipophilic xenobiotics, which enhances solubility and rapid elimination of the molecule.

US Food and Drug Administration (FDA): A US federal agency responsible for protecting public health by assuring the safety and efficacy of drugs, biological products, medical devices, and other products.

Xenobiotic: A substance foreign to the body.

Key Abbreviations

ANOVA	Analysis of Variance
CPNDS	Canadian Pharmacogenomics Network for Drug Safety
CNS	Central Nervous System
CEO	Chief Executive Officer
CSO	Chief Scientific Officer
CDSS	Clinical Decision Support Systems
CPIC	Clinical Pharmacogenetics Implementation Consortium
cDNA	Complementary DNA
DPWG;	Dutch Pharmacogenetics Working Group
ESI	Ethical Legal and Societal Issues
RNPGx	French National Network of Pharmacogenetics
GWAS	Genome-Wide Association Studies
HIV	Human Immunodeficiency Virus
I-PWG	Industry Pharmacogenomics Working Group
KAP	Knowledge, Attitudes, and Practices
NGS	Next Generation Sequencing
NOD	Non-Obese Diabetic
PCR-RFLP	PCR-Restriction Fragment Length Polymorphism
PM	Poor Metabolizer
R&D	Research and Development
RT-PCR	Reverse Transcription-PCR
VUS	Variants of Unknown Clinical Significance
WHO	World Health Organization

References

Ampadu HH, Hoekman J, de Bruin ML, Pal SN, Olsson S, Sartori D, Leufkens HGM, Dodoo ANO (2016) Adverse drug reaction reporting in Africa and a comparison of individual case safety report characteristics between Africa and the rest of the world: Analyses of spontaneous reports in vigibase®. Drug Saf 39:335–345

Arbitrio M, Scionti F, Di Martino MT, Caracciolo D, Pensabene L, Tassone P, Tagliaferri P (2021) Pharmacogenomics biomarker discovery and validation for translation in clinical practice. Clin Transl Sci 14:113–119

Begley CG (2013) Six red flags for suspect work. Nature 497:433–434

Begley CG, Ellis LM (2012) Raise standards for preclinical cancer research. Nature 483:531–533

Begley CG, Snapinn SM (2021) Comment on "the role of p-values in judging the strength of evidence and realistic replication expectations". Stat Biopharm Res 13:40–42

Bienfait K, Chhibber A, Marshall J-C, Armstrong M, Cox C, Shaw PM, Paulding C (2021) Current challenges and opportunities for pharmacogenomics: perspective of the Industry Pharmacogenomics Working Group (I-PWG). Hum Genet 141:1165. https://doi.org/10.1007/s00439-021-02282-3

Biologics Market – Growth, Trends, COVID-19 Impact, and Forecasts (2022–2027). In: ReportLinker. https://www.reportlinker.com/p06283186/Biologics-Market-Growth-Trends-COVID-19-Impact-and-Forecasts.html?utm_source=GNW. Accessed 27 Jun 2022

Center for Drug Evaluation and Research (CDER) (2021) About biomarkers and qualification. U.S. Food and Drug Administration. https://www.fda.gov/drugs/biomarker-qualification-program/about-biomarkers-and-qualification. Accessed 5 Apr 2022

Dere WH, Suto TS (2009) The role of pharmacogenetics and pharmacogenomics in improving translational medicine. Clin Cases Miner Bone Metab 6:13–16

Ekroos M, Sjögren T (2006) Structural basis for ligand promiscuity in cytochrome P450 3A4. Proc Natl Acad Sci 103:13682–13687

Errington TM, Mathur M, Soderberg CK, Denis A, Perfito N, Iorns E, Nosek BA (2021a) Investigating the replicability of preclinical cancer biology. elife 10:e71601

Errington TM, Denis A, Perfito N, Iorns E, Nosek BA (2021b) Reproducibility in cancer biology: challenges for assessing replicability in preclinical cancer biology. elife 10:e67995

Fell MJ, Mirescu C, Basu K et al (2015) MLi-2, a potent, selective, and centrally active compound for exploring the therapeutic potential and safety of LRRK2 kinase inhibition. J Pharmacol Exp Ther 355:397–409

Guan L, Yang H, Cai Y, Sun L, Di P, Li W, Liu G, Tang Y (2019) ADMET-score – a comprehensive scoring function for evaluation of chemical drug-likeness. MedChemComm 10:148–157

King C, Sarvetnick N (2011) The incidence of type-1 Diabetes in NOD mice is modulated by restricted flora not germ-Free conditions. PLoS One 6:e17049

Kneller R (2010) The importance of new companies for drug discovery: origins of a decade of new drugs. Nat Rev Drug Discov 9:867–882

Masimirembwa C, Dandara CC, Leutscher PD (2016) Rolling out efavirenz for HIV precision medicine in Africa: are we ready for pharmacovigilance and tackling neuropsychiatric adverse effects? OMICS J Integr Biol 20:575–580

Mbavha BT, Kanji CR, Stadler N, Stingl J, Stanglmair A, Scholl C, Wekwete W, Masimirembwa C (2022) Population genetic polymorphisms of pharmacogenes in Zimbabwe, a potential guide for the safe and efficacious use of medicines in people of African ancestry. Pharmacogenet Genomics 32:173. https://doi.org/10.1097/FPC.0000000000000467

Mendel DB, Cherrington JM, Laird AD (2015) CCR 20th anniversary commentary: determining a pharmacokinetic/pharmacodynamic relationship for sunitinib—a look back. Clin Cancer Res 21:2415–2417

Peers IS, Ceuppens PR, Harbron C (2012) In search of preclinical robustness. Nat Rev Drug Discov 11:733–734

Prinz F, Schlange T, Asadullah K (2011) Believe it or not: how much can we rely on published data on potential drug targets? Nat Rev Drug Discov 10:712

Radouani F, Zass L, Hamdi Y et al (2020) A review of clinical pharmacogenetics Studies in African populations. Pers Med 17:155–170

Roosan D, Hwang A, Roosan MR (2021) Pharmacogenomics cascade testing (PhaCT): a novel approach for preemptive pharmacogenomics testing to optimize medication therapy. Pharmacogenomics J 21:1–7

Rydzewski RM (2008) Chapter 9: ADME and PK properties. In: Real world drug discovery: a chemist's guide to biotech and pharmaceutical research. Elsevier, Chem, pp 353–429

Shek D, Read SA, Ahlenstiel G, Piatkov I (2019) Pharmacogenetics of anticancer monoclonal antibodies. Cancer Drug Resist 2:69–81

U.S. Food and Drug Administration (2005) Center for Drug Evaluation and Research (CDER). Guidance for industry: estimating the maximum safe starting dose in initial clinical trials for therapeutics in adult healthy volunteers, July 2005. Available at: https://www.fda.gov/media/72309/download

van der Wouden CH, Böhringer S, Cecchin E et al (2020) Generating evidence for precision medicine: considerations made by the Ubiquitous Pharmacogenomics Consortium when designing and operationalizing the PREPARE study. Pharmacogenet Genomics 30:131–144

Weitzel KW, Alexander M, Bernhardt BA et al (2016) The IGNITE network: a model for genomic medicine implementation and research. BMC Med Genet 9:1

World Health Organization (2005) Annex 1. WHO guidelines on nonclinical evaluation of vaccines, Technical report series, no. 927. World Health Organization, Geneva, pp 31–63

Preparing for the Clinic

Daria Mochly-Rosen, Kevin Grimes, Carol D. Karp,
Robert Lum, Mark Backer, Lyn Frumkin,
and Ted McCluskey

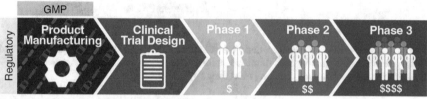

Daria Mochly-Rosen, Ed, Kevin Grimes, Ed.

Transitioning from preclinical to clinical development can be both an exhilarating and a sobering experience. At this point, our drug has passed through a rigorous battery of preclinical testing and appears to have the desired safety profile and pharmacologic characteristics for advancement to human study. However, embarking upon human experimentation is a serious undertaking. We must ensure that our clinical trial is designed and conducted in as safe a manner as possible. We have an ethical responsibility to ensure that our clinical trial has been designed to optimize the probability of obtaining meaningful results as we are exposing human subjects to a potentially toxic new molecular entity. Furthermore, the drug must be manufactured and quality-tested using exacting standards, typically by a reputable contract manufacturing

D. Mochly-Rosen (✉) · K. Grimes
Chemical and Systems Biology, Stanford University School of Medicine, Stanford, CA, US
e-mail: sparkmed@stanford.edu; kgrimes@stanford.edu

C. D. Karp
Prothena Biosciences, Brisbane, CA, US

R. Lum
Concentric Analgesics Inc., San Francisco, CA, US

M. Backer
Alava Biopharm Partners LLC, San Mateo, CA, US

L. Frumkin · T. McCluskey
SPARK at Stanford Advisor, Stanford, CA, US

© The Author(s), under exclusive license to Springer Nature Switzerland AG 2023
D. Mochly-Rosen, K. Grimes (eds.), *A Practical Guide to Drug Development in Academia*, https://doi.org/10.1007/978-3-031-34724-5_4

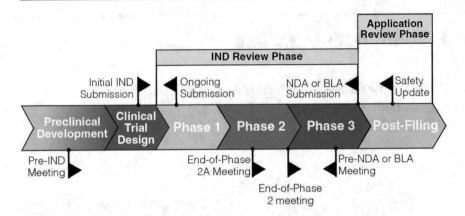

Fig. 4.1 Industry-FDA interactions during development. (Adapted from FDA (U.S. Food and Drug Administration (FDA) 2004)). Abbreviations: *BLA* Biologics License Application, *IND* Investigational New Drug Application , *NDA* New Drug Application

organization (CMO). An Institutional Review Board (IRB) approval (or Ethics Committee, outside of the US) must be obtained to ensure that patients are protected and that the risk-to-benefit ratio is acceptable for any clinical study.

Early in the planning process, the development team should engage experts in drug manufacturing and quality control (QC), clinical trial design, and regulatory science. Before the clinical trial can commence, the sponsor (company or physician) must file an Investigational New Drug application (IND) with the US Food and Drug Administration (FDA) or the equivalent authorities in other jurisdictions. The IND provides detailed information on the nonclinical pharmacology and safety studies, the drug manufacturing and quality assurance (QA) process, and the clinical trial protocol. If the submitted IND is approved by FDA, subsequent meetings will take place between the development team and FDA until the submission of a New Drug Application (NDA) or Biologic License Application (BLA). Key FDA-industry interactions are summarized in Fig. 4.1.

This chapter provides an overview of clinical trial design, acquisition of clinical grade drug product (DP), and regulatory considerations during the transition into the clinic. Because clinical trials are so expensive, failure to appropriately design the clinical trial can obscure efficacy and force a program to close. Even seasoned clinicians would be wise to talk with other experts before embarking on a clinical trial.

4.1   Regulatory Considerations in Product Development

Carol D. Karp

When determining the regulatory pathway for development of a new product, it is essential to consider the overall benefit-risk profile, target disease, patient population, intended use, and claims for the product. Generally, the broader the target population or product claims, the greater the burden of evidence that will be required to support product approval. The optimal regulatory approach for a specific product should be adapted based upon these key considerations.

Another important regulatory consideration is whether the selected clinical indication meets the definition of a rare disease. Under the Orphan Drug Act, FDA can grant orphan drug designation to a therapeutic being developed for a rare disease, which is defined as a condition affecting fewer than 200,000 individuals in the US. Orphan drug designation qualifies the sponsor for incentives that include tax credits for qualified clinical trials, exemption from paying Prescription Drug User Fee Act (PDUFA) user fees, and seven years of market exclusivity following approval. The sponsor must submit a separate request for orphan drug designation to the agency.

In the US, the Prescription Drug User Fee Act (PDUFA), originally enacted in 1992, enables FDA to collect fees from sponsors submitting applications to market human drug and biological products. PDUFA established goals and procedures intended to facilitate timely access to safe, effective, and innovative new medicines for patients. PDUFA VII, enacted in 2022, includes reauthorization of the Act through fiscal years 2023–2027. The 2023 fee for submission of an NDA or BLA to FDA under PDUFA VII is $3,242,026. An NDA requests permission to market a new drug product in the US, while a BLA requests permission to introduce a new biologic product to the market; biologic products are discussed in greater detail in Sect. 4.3.

4.1.1 Investigational New Drug Application (IND)

The first step to initiate clinical studies for a new product is the submission of an IND to FDA. Key components of an IND consist of the following: (1) nonclinical pharmacology and toxicology summaries and study reports to demonstrate that the investigational drug is reasonably safe for initial testing in human subjects. When repurposing a previously approved product, this may entail support for testing of a new dose level, formulation, route of administration, patient population or target disease; and (2) quality information, referred to as chemistry, manufacturing, and controls (CMC), to assure product safety, identity, quality, purity, and strength. This includes descriptions of the composition, source, manufacturer, testing, specifications and stability of the drug substance, or active pharmaceutical ingredient (API), and the finished drug product (DP); (3) clinical information, including the proposed clinical study protocol, data from previously conducted clinical studies, if available, and identification of the clinical investigator(s), to assure that subjects will not be exposed to unnecessary risks and to confirm the qualifications of the investigator(s); (4) investigator's brochure, which provides a compilation of available clinical and nonclinical data and guidance on the study rationale, dosing regimen, and clinical management of study subjects, including safety monitoring; and (5) the general investigational plan for clinical studies to be conducted over the next year.

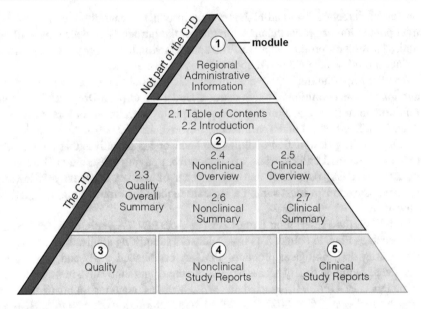

Fig. 4.2 IND organization (CTD format). (Adapted from ICH, https://www.ich.org/page/ctd).
Abbreviations: *CTD* Common Technical Document, *ICH* International Council for Harmonisation
of Technical Requirements for Pharmaceuticals for Human Use

An IND is submitted to FDA as a Common Technical Document (CTD), the
accepted format for both INDs and NDAs/BLAs and for equivalent submissions in
the European Union (EU), Japan, and other countries, such as Canada and
Switzerland (Fig. 4.2). The CTD format allows the IND to serve as the initial building blocks for what will eventually become an NDA or BLA to obtain FDA approval
to market the product upon successful completion of a clinical program. The initial
IND review period is a 30-day process that encompasses assessments by FDA medical, pharmacology/toxicology, and chemistry reviewers to determine if the initial
clinical study can be safely conducted.

Typically, an IND is submitted by a pharmaceutical or biotech company sponsor.
Alternatively, in an academic setting, a physician may submit an Investigator
IND. In this case, the sponsoring physician both initiates and conducts the clinical
investigation, and the investigational drug must be administered or dispensed under
the direction of the sponsoring physician. Investigator INDs are exempt from eCTD
requirements, and the IND may therefore be submitted in paper format, although
FDA both encourages and prefers submissions in an alternative electronic format
such as PDF files following the CTD structure.

4.1.2   IND Considerations

Exemptions from the IND requirements may apply to clinical studies of approved
DPs that do not involve new routes of administration, dosing levels, patient

populations, or other factors that might significantly increase the risks associated with the approved commercial use of the product.

In addition, FDA has made available a number of expedited approaches for development and review of new DPs. An exploratory IND can allow for screening of multiple closely related active moieties to identify the preferred compound or formulation to be taken forward for development under a formal IND. This approach is intended to provide greater flexibility for early phase 1 screening or microdose studies to assess attributes such as mechanism of action or pharmacokinetics. Given the limited human exposure, such studies can be initiated with a less extensive pre-clinical program than that required for a traditional IND.

Fast track is an FDA program spanning both the IND and NDA (or BLA) development stages for investigational products with the potential to address (1) a serious or life-threatening condition and (2) an unmet medical need. Fast track designation may be requested with an initial IND or at any point prior to NDA submission. Fast track is an overarching regulatory approach that can include the following: (1) close ongoing communication with FDA from pre-IND through pre-NDA meetings, (2) rolling review of sections of an application by FDA in advance of the complete NDA/BLA submission, and (3) eligibility for priority review, if supported by clinical data at the time of NDA submission, which allows for a shorter FDA review period for the application.

In 2012, FDA established Breakthrough Therapy designation for drugs intended to treat a serious or life-threatening disease and with preliminary clinical evidence suggesting that a substantial improvement in one or more relevant clinical endpoints over current treatment options may be provided. Breakthrough Therapy designation provides all the features of fast track status, together with intensive FDA guidance involving senior managers and experienced reviewers, to support the conduct of efficient clinical trials and to expedite the overall development program.

4.1.3 Meetings with FDA

Meetings between FDA and a sponsor or applicant of an IND or NDA are not required but are often encouraged to obtain FDA guidance at critical points during the drug development process. A pre-IND meeting provides an opportunity for a sponsor to introduce the investigational product and the rationale for the development program. The meeting can be held as a face-to-face meeting, teleconference, or as written response only (WRO).

The pre-IND meeting request, typically a 3–5 page document, includes the product name, chemical structure, proposed indication, objectives of the meeting, proposed agenda, list of specific questions, list of participants for the sponsor, and requested FDA participants. A background package of about 50–100 pages is submitted in advance of the meeting and includes (1) an introduction of the product concept, (2) the proposed indication, (3) an overview of the development plans, (4) a summary of the pharmacology and toxicology studies that have been conducted to date, (5) initial plans for additional nonclinical studies, (6) a summary of CMC information available on the drug substance and DP, (7) the initial clinical development plans, (8) a synopsis of the initial clinical study protocol, and (9) specific questions or issues that require FDA's feedback (Box 4.1).

Box 4.1: Pre-IND Meeting Timeframes
Day 0: Submit meeting request
Day 21: FDA response to request
Day ≤30: Submit briefing package
Day ≤60: Pre-IND meeting
Day ≤90: FDA meeting minutes

Three types of meetings occur between FDA and a sponsor or applicant. Pre-IND meetings are classified as Type B meetings; end-of-phase 2, prephase 3, and pre-NDA meetings are also classified as Type B meetings. Type A meetings are generally reserved for situations such as a clinical hold, when FDA has determined that a clinical study cannot be initiated or continued. Type C meetings are those not covered by Type A or Type B.

Box 4.2: Timeframes from FDA Receipt of Written Request to Meeting
Type A: 30 Days
Type B: 60 Days
Type B (End-of-Phase): 70 Days
Type C: 75 Days

Box 4.3: What Surprised an Academic?
When I attended a pre-IND meeting with KAI's team and the regulatory consultant, her last words before we entered the room were "keep quiet and act pretty!" There was no time for me to express my outrage, but the gist of her instruction (although not most appropriately stated) was correct.

When academic inventors attend an FDA meeting, they should be well prepared and rehearsed. It is important that the meeting is led by the regulatory expert. In some cases, the academic inventor may be asked questions by FDA officers. In that case, they should focus on responding carefully, but briefly. It is important to remember that FDA meetings are not open-ended scientific discussions. The time with FDA is precious and should be used only to address prespecified questions that will support the clinical trial. —DM-R

4.1.4 New Drug/Biologics License Applications

An NDA or BLA, the application to obtain FDA approval to market a new drug or biologic product in the US, is based on the nonclinical, clinical, and CMC data generated during the IND development stage. The data provided in an NDA or BLA enables FDA to determine whether (1) the product is safe and effective for its proposed use and the benefits outweigh the risks, (2) the proposed product labeling is appropriate and adequate, and (3) the manufacturing methods and controls to maintain product quality are sufficient.

As noted above, FDA may grant priority review for an NDA or BLA if the product is intended to treat a serious condition and would provide a significant improvement in safety or efficacy, reducing the total review period from 12 months to 8 months for innovative products (new molecular entities). In addition, the accelerated approval pathway is available for a product that treats a serious condition and provides a meaningful advantage over available therapies, with FDA approval based upon a demonstrated effect on a surrogate endpoint that is reasonably likely to predict clinical benefit.

For a product intended to provide chronic treatment for a non-life-threatening condition, the extent of population exposure to support approval is approximately 1500 patients, with 100 patients treated for 12 months and 300–600 patients treated for 6 months. For each new product, the regulatory pathway and overall burden of evidence should be considered in the context of the target disease, patient population, product claims, and overall benefit/risk profile.

Box 4.4: What Surprised an Academic?

For academics who have not previously interacted with FDA, the depth of questions and data requests, particularly pertaining to the manufacture and testing of a drug substance and DP, can come as a surprise, as can the nature and extent of requests pertaining to a clinical protocol. In addition, the experiences of academic colleagues developing products for different diseases and patient populations at the same institution may vary substantially in terms of the feedback they receive from FDA. It is important to keep in mind that regulatory requirements can be dependent upon the specific product and target clinical use, and that FDA, rather than being a monolith, is composed of various Offices, Divisions and reviewers. —*CDK*

The Bottom Line

FDA performs a critical role in ensuring that new therapies are both safe and effective. A collegial working relationship with the agency and familiarity with applicable regulations and guidelines can facilitate the transition from lab bench to clinic and increase the probability of success for a development program.

Resources
- FDA Regulations. IND Content and Format: https://www.accessdata.fda. gov/scripts/cdrh/cfdocs/cfcfr/CFRSearch.cfm?fr=312.23.
- FDA Guidance for Industry. Organization of the CTD: http://www.fda. gov/downloads/Drugs/GuidanceComplianceRegulatoryInformation/ Guidances/UCM073257.pdf.
- Guidance for Industry. IND Applications Prepared and Submitted by Sponsor-Investigators. May 2015: http://www.fda.gov/ucm/groups/ fdagov-public/@fdagov-drugs-gen/documents/document/ucm446695.pdf.
- FDA Guidance for Industry. INDs. Determining whether Human Research Studies can be conducted without an IND: http://www.fda.gov/downloads/ Drugs/GuidanceComplianceRegulatoryInformation/Guidances/ UCM229175.pdf.
- FDA Guidance for Industry, Investigators and Reviewers: Exploratory IND Studies: http://www.fda.gov/downloads/Drugs/GuidanceCompliance RegulatoryInformation/Guidances/UCM078933.pdf.
- FDA Draft Guidance for Industry: Expedited Programs for Serious Conditions—Drugs and Biologics: https://www.fda.gov/regulatory- information/search-fda-guidance-documents/expedited-programs-serious- conditions-drugs-and-biologics.
- FDA Guidance for Industry: Formal Meetings between the FDA and Sponsors or Applicants: https://www.fda.gov/regulatory-information/ search-fda-guidance-documents/formal-meetings-between- fda-and-sponsors-or-applicants-pdufa-products-guidance-industry.
- ICH Guideline for Industry: Extent of Exposure to Assess Clinical Safety: http://www.fda.gov/downloads/Drugs/GuidanceComplianceRegulatory Information/Guidances/UCM073083.pdf.

4.2   Manufacturing and Quality Control

Robert Lum

4.2.1   Regulatory Considerations

The manufacturing of drugs is regulated by FDA in the US, by the European Medicines Agency (EMA) in the EU, by the Pharmaceuticals and Medical Devices Agency (PMDA) in Japan, and by various other regulatory agencies throughout the world. The International Council for Harmonisation of Technical Requirements for Pharmaceuticals for Human Use (ICH) represents an ongoing effort by regulatory authorities and pharmaceutical trade organizations of the US, Europe, and Japan to harmonize drug regulatory requirements.

Nonetheless, regulatory requirements for clinical drug supplies can and do vary by geographic region. When planning clinical studies, the drug supply must meet the regulatory requirements for each country in which the study will be conducted. While national regulatory agencies and ICH have published guidance documents, discussions with individual regulatory agencies can also be helpful during the planning process.

If studies are planned in the US only, the products will need to be approved and released by a QA group independent of the manufacturing entity. If studies are planned in the EU, the products will need to be released by a licensed "Qualified Person" (QP). These quality groups must certify that the products meet predetermined specifications, were manufactured according to the applicable "Good Manufacturing Practice" (GMP), and are suitable for human use.

4.2.2   Manufacturing Requirements

Facilities that manufacture drugs for clinical trials must operate in compliance with applicable GMP. US drug GMP regulations are published in the Code of Federal Regulations (CFR) under Title 21, Parts 210 and 211 "Good Manufacturing Practice." These regulations require documentation of all aspects of the manufacturing process.

Key points in the GMP regulations include the following:

- There must be an independent quality function.
- All materials used in manufacturing must be traceable to the original manufacturer and lot number.
- Inventories must be properly stored and maintained, which includes documenting the suitability of all raw materials, solvents, and components for GMP manufacturing.
- There must be documented evidence that operators are educated and have current training in all functions needed to perform their GMP duties.
- All equipment used must be calibrated using National Institute of Standards and Technology (NIST) traceable standards.
- The manufacturing steps and testing methods must be written and approved by management and QA prior to execution in manufacturing. This is typically called a Master Batch Record (MBR).
- Each manufacturing step performed must be documented and critical steps must also be witnessed, as evidenced by dated signature. A completed MBR is often referred to as an Executed Batch Record (EBR).
- Any deviations to the written methods must be documented, investigated, and approved by QA.
- Specifications for incoming raw materials, components, intermediates, and final product must be written and approved by QA prior to testing.

- The product must meet predetermined specifications.
- Quality systems including Standard Operating Procedures (SOP), internal audits, Change Controls, Corrective and Preventive Actions (CAPA), scheduled management reviews, and a quality manual must be in place.

In addition to these basic requirements, there are many additional regulations specific to biologics, peptides, oligonucleotides, sterile products, cells, devices, botanicals, extended-release formulations, etc. It is essential to have someone familiar with GMP manufacturing for your type of product and the applicable regulations.

> **Box 4.5: What Surprised an Academic?**
> Early in our phase 2a clinical trial with the first compound, our VP of Drug Development decided to manufacture a new batch of GMP material in spite of the fact that we had plenty made for the trial size. This was a huge expense and I pushed back on it, arguing that it was wasteful. When our first batch failed specifications with unacceptable levels of degradation due to the slow enrollment of our trial, it took the company 2 days to retrieve the old drug from the sites and switch to the new batch. Planning for failure and building contingencies were critical for the success of the trial. —*DM-R*

4.2.3   Testing Requirements

Drugs used in clinical studies must be tested to assure that they meet preapproved specifications each for API and DP. This testing typically includes some measure of potency and assays for impurities and degradation products. If the API has chiral centers, testing for enantiomer purity may be warranted. Depending on the dosage form of the DP, safety tests for sterility, bioburden, endotoxins, and heavy metals may also be required.

Assays used to test products must be "validated"; validation is a formal analytical process demonstrating that the assay performs as intended and is specific, reproducible, and precise. Assays used in early clinical trials must be documented to be at least scientifically sound—this is often referred to as "qualified." The use of these terms varies significantly from company to company; make sure to ask exactly what is meant.

4.2.4   Stability Testing

There must be data available to assure that the API and DP used in clinical testing are stable throughout the duration of the trial. If this is an existing product for a new use, such data may already exist. However, stability testing must be conducted for a new formulation or drug product dosage. A stability-indicating assay is developed and qualified to detect impurities and degradation products. Forced degradation studies attempt to accelerate degradation by exposure to stressful storage conditions, usually excess pH, heat, moisture, or light. In many cases, a high-performance liquid chromatography (HPLC) assay is included to detect degradation products.

For new IND submissions, it is typical to submit 3–6 months of stability data on GMP API and 3 months of data for GMP DP. "Supportive" stability studies that were performed on research material prior to GMP manufacturing can also be included. Stability testing of the API and DP must continue during the clinical trial period. Typically, stability testing is performed at three different temperature/relative humidity conditions and at time 0, 1, 3, 6, 9, 12, 18, 24, and 36 months, as described in the ICH Stability Guidelines. Quality control must monitor the stability results in real time and if a product fails to meet specifications, the product must be removed from clinical use.

Box 4.6: What Surprised an Academic?

I was surprised one afternoon to find hundreds of drug vials on the precious lab space that I secured for our new hire at KAI. When told that they were running a stability study, I was further irritated; I had already performed an HPLC run of a compound that we saved in the fridge for a few months in my lab at Stanford University. Why repeat it and why in my precious lab space?! The reason is that the experimental rigor and degree of documentation for this study far exceeded what I had thought was necessary. —DM-R

4.2.5 Selecting a Contract Manufacturing Organization

There are several important questions when selecting a CMO:

1. Does it have quality systems for GMP manufacturing processes in place?
2. Does it have experience in developing and manufacturing GMP API or DP that meets your requirements?
3. Does it have the equipment and controls to handle your specific project? Many cancer drugs are considered "high potent APIs" and require specialized handling.

You absolutely do not want the CMO learning GMP basics on your project. You are paying for both technical expertise and regulatory compliance. You should find an expert advisor that is familiar with GMP manufacturing to help identify appropriate CMOs. It is recommended to prepare a "Request for Proposal" (RFP) that you can send to potential companies, indicating what services you need such as manufacturing, analytical development, QC testing, stability testing, packaging, and labeling. Be sure to include specifics about your desired product. For example, specify the quantity of API and what specifications it must meet, or number of bottles of finished drug product. Provide details including availability of starting materials such as API, stability study outline, and timeframe for the project. Request a detailed quote with the pricing broken down for each activity. Once the company has carefully read your RFP, a teleconference can be a helpful next step. After you have compared quotes, you will typically need to go back to the companies for clarifications. It is advisable to visit the CMO site in person with your expert advisor.

4.2.6 CMC Section of the IND

The CMC section of the IND must include a detailed description of the manufacturing and testing of the API and DP. This section of the filing will also include an overview of the manufacturing process, specifications, test methods, QC analysis, and stability information for each product lot intended for the clinic. A list of all facilities involved in the manufacturing, testing, holding, and distribution of the product is also required. Occasionally, you may be performing clinical testing using product manufactured by a company that holds a "Master File" with FDA from previous filings. Under such circumstances, the CMC section can be replaced by a reference to the "Master File."

> **The Bottom Line**
> Although the regulatory requirements for manufacturing and quality assurance seem rather elaborate, they exist to protect both patients and drug manufacturers. Consistency of DP is essential for reproducible clinical studies that demonstrate both safety and efficacy. These standards underscore a critically important issue: Aggressive quality is not only good science, but helps ensure the safety of the experimental subjects (the patients).

4.3 Technical Development and Manufacturing of Biological Products

Mark Backer

The development of biological products, or "large molecules," has important differences from drugs created by synthetic chemistry. The preparations to support a proposed clinical testing plan must address the same categories as small molecules, as outlined in IND guidance documents, but the nature of these activities can be substantially different based on the production methods and product characteristics associated with biologicals. An overview of early-stage biological product development is provided here, with a focus on manufacturing, testing, and regulatory expectations.

The technical and regulatory foundation for modern biological products was established by the development of life-saving vaccines and protein replacement therapies, such as bovine insulin. These products were derived from animals, and regulators quickly learned that the benefits of these products were balanced by some new complexities and risks due to the potential for variability or unwanted contaminants from the biological source. With the advent of eggs and then aseptic cell culture for vaccine production, followed by recombinant DNA (rDNA) technology in the late 1970s and monoclonal antibody technology in the 1980s, it became possible to produce a much wider variety of biological products with improved control over product consistency and safety. Still, the biological origin and the natural heterogeneity of these products require a different development approach than small molecules.

In the early days of biological products, a guiding philosophy was "the process is the product"; in other words, process changes were discouraged because the resulting product could not be characterized well. Potential effects on product quality and safety could not be easily assessed without repeating clinical trials. Now, a framework is in place to support process evolution and biosimilar (off-patent) products, but this depends on a thorough comparability analysis of drug characteristics.

Biological products include a wide variety of product types; at one extreme, some small peptides, proteins, and nucleic acid products can in fact be manufactured by nonbiological synthesis. At the other extreme, live viruses and whole cells have been developed as commercial products. The development approach needs to be adapted for each type of product. As an introduction, this discussion will focus on recombinant protein products and monoclonal antibodies, which are the most common biological products.

4.3.1   Expression Systems

The first biological products that used rDNA technology were produced in *E. coli* bacteria, based on 1972 technology developed by Boyer and Cohen from the University of California San Francisco and Stanford University, respectively (Cohen et al. 1973; Khan et al. 2016). This technology was used to produce human insulin and growth hormone, for example. These products established the method of diverting the *E. coli* biological machinery to manufacture the protein of interest using a plasmid that includes a promoter driving expression of a DNA transgene encoding the target protein. Microbial expression systems are popular because they can generate large amounts of recombinant protein thanks to strong translation machinery and dense culture conditions. There are some limitations, however. Large mammalian proteins that feature numerous disulfide cross-links, glycosylation, or post-translational modifications generally cannot be manufactured in *E. coli* or yeast. In addition, proper protein folding in microbial systems can be a problem, and the target protein must often be refolded after recovery from misfolded, aggregated material called inclusion bodies.

For more complex proteins and antibodies, mammalian cell culture is used for manufacturing. The first antibody products were produced, based on the 1975 technology of Kohler and Milstein at the MRC Laboratory of Molecular Biology in Cambridge, UK (Köhler and Milstein 1975), by culturing hybridoma cell lines, which are created by fusing a continuous B cell line (myeloma) with another B cell that produces the antibody of interest. Today, the generation of a hybridoma is generally followed by transferring the antibody heavy and light chain genes to a different cell line with promoters optimized for manufacturing: typically, Chinese hamster ovary (CHO) cells. The human embryonic kidney (HEK) 293 cell line is also used in manufacturing recombinant proteins, and other human and animal cell lines are used to produce viral-based vaccines. A panel of candidate recombinant clones (each clone grown from a single cell) is tested for productivity and quality, and the chosen clone is expanded to create a research cell bank that serves as the basis for further testing and eventually manufacturing.

4.3.2   Preparing to Manufacture Products for Clinical Testing

Once an antibody or protein product has been tested in animal models and there is a rationale for further development, preparations for clinical testing begin. FDA has provided guidance that outlines the expectations for sponsors' documentation and testing when using antibodies, and it is prudent to review this before making the hybridoma cell line. As an early step, the research bank is used to create a Master Cell Bank (MCB) that establishes the foundation for consistent production under GMP conditions. For pivotal trials and commercial manufacturing, a two-tiered bank system is required, and a Working Cell Bank (WCB) is also produced and tested. Some sponsors produce both the MCB and WCB prior to phase 1 testing to eliminate any subsequent product comparability issues, but WCB production is often deferred to save time and money for first-in-human trials. If the manufacturing process features a virus for production of vaccine antigen or use as a genetic vector, a similar approach to banking is used (preparation of a Master Virus Seed and Working Virus Seed in addition to the MCB and WCB for the cell substrate).

4.3.3   Upstream and Downstream

Manufacturing proceeds by expanding the MCB clone from the seed stock in a series of increasingly larger vessels, through the production bioreactor scale, using customized growth media in axenic culture (sterile except for the production cell line itself). Bioreactors provide tight control over oxygen levels, pH, and mixing of the culture; and concentrated nutrients are added to increase cell density. Mammalian cell culture is practiced in industry at a scale up to 20,000 L of bioreactor volume, but production for phase 1 testing is conducted at a smaller scale—typically 100–1000 L depending on the trial design and dosage. In general, fed-batch cell culture densities are in the range of $5–10 \times 10^6$ cells/mL and nonoptimized antibody titers are in the range of 1–2 g/L of bioreactor volume. For products with anticipated large demand, further optimization is usually performed prior to pivotal trials to reach titers above 4 g/L. Antibodies and recombinant proteins are excreted from the producing cells, so the product is harvested by removing the cells using a centrifuge or filter. The process steps through this stage are called the upstream process.

The cell-free harvest must be purified and concentrated to produce a composition suitable for use in humans. These downstream process steps generally include an affinity purification step for antibodies (using a Protein A resin for capture) and at least two column chromatography steps to separate the target protein from impurities and unwanted variants (such as aggregates, clipped proteins, and dimers). Polymer membranes that are permeable only to water and small molecular weight molecules are used for concentration and buffer exchange; these steps are called ultrafiltration (UF) and diafiltration (DF). The resulting material, which should be close to the intended clinical formulation, is called drug substance (DS), which has the same meaning as API, the term typically used for small molecules.

Because of the chance that a virus could be present in the producing cell line or introduced in the upstream process, extra downstream steps are performed to reduce this risk (e.g., a low pH step and nanofiltration using a filter that would remove virus). Specific viral clearance studies are performed to demonstrate that the steps effectively remove a panel of representative virus types. These studies, like testing for adventitious agents, are generally contracted out to one of a small number of industry vendors with the required capabilities and experience.

Many antibody and recombinant protein projects today are efficiently supported by outsourcing the development effort described above to a Contract Development and Manufacturing Organization (CDMO), rather than assembling the capabilities for "in-house" development and manufacturing under suitable GMP. A CDMO may apply somewhat generic methods for development and manufacturing, or it may offer a tested platform approach for the producing cell line, expression system, media components, etc. These platform technologies may be offered without further obligation as part of the service contract, but in some cases proprietary methods are used that may carry fees or royalties, or restrictions on technology transfer to a different manufacturer.

4.3.4   Drug Product

Most biological products are sterile and are administered by a parenteral route (intravenous or injection). The drug substance is diluted or reformulated as needed, passed through a sterilizing filter, and then filled into vials or prefilled syringes in special clean rooms to maintain sterility. They are then labeled and packaged into the format needed for clinical testing. Unlike some small molecule drugs, proteins cannot be sterilized by heat or radiation, so maintaining sterility through the filling process is critical for patient safety. If the filling vendor has not recently used the same components (vials and stoppers) at a similar scale for GMP manufacturing, filling sterility must be assured in advance by performing a media fill (a dummy run using sterile media); this time and expense needs to be considered in the project plan. Biologicals may be lyophilized and reconstituted for administration (potentially a challenging development activity), but most are stored as refrigerated or frozen liquids, particularly in early development. Figure 4.3 illustrates the CMC path to licensure for a biological product.

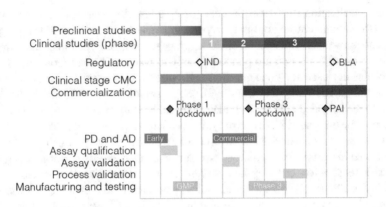

Fig. 4.3 CMC view of the path to licensure for a biological product. A typical project may take around 9 years from initiating development of a candidate to BLA filing; this Gantt chart outlines important development steps. Abbreviations: *AD* Assay Development, *BLA* Biologics License Application, *CMC* Chemistry, Manufacturing, and Controls, *GMP* Good Manufacturing Practice, *IND* Investigational New Drug Application, *PAI* Pre-Approval Inspection, *PD* Process Development

4.3.5 Testing and Compliance

A successful IND requires performing extensive safety testing and characterization following GMP regulations. The term "CMC" has become an umbrella term that refers to the process development and analytical development work supporting product development, together with Manufacturing and Quality activities—rather than simply the CMC section of the IND itself (Tables 4.1 and 4.2).

There are some key considerations for biological material CMC:

- Documentation is critical. Even the construction of the Research Cell Bank should be thoroughly documented, such as the origin, lot numbers, and qualification (if any) of materials that are used (gene sources or synthesis, creation of the expression construct itself, cell lines, serum, and any material of animal origin). If serum must be used at any point in development, it should be obtained from a transmissible spongiform encephalopathy (TSE)-free country such as New Zealand and purchasing records and lot numbers retained.
- Cell lines used in the biologic construction or manufacturing should have documented origin and freedom to operate (e.g., a research license from ATCC). Clinical use of particular promoters, genetic elements, or humanization methods may also trigger licensing fees or unwanted royalty obligations attached to the product, although sometimes these charges are an acceptable burden in return for enabling the best technical approach.
- Cell banks and drug substance batches need to be thoroughly tested for adventitious agents (such as unwanted viruses or bacteria), and the recombinant construct should be sequenced to assure that the correct protein sequence is produced and no extraneous sequences are present.

Table 4.1 Typical specifications for quality testing results of drug substance and drug product of a monoclonal antibody prior to entry into a phase 1 clinical trial

Drug substance in-process tests (performed on clarified harvest where detection is easiest)		
Test	*Category*	*Specification*
Bioburden (culture)	Purity	$\leq$5 cfu/mL
Protein A content (ELISA)	Purity	<20 ppm
CHO host cell protein content (ELISA kit)	Purity	<50 ppm
CHO host cell DNA content (PCR)	Purity	10 pg/mg
Mycoplasma detection (culture)	Safety	None detected
Mycobacterium tuberculosis	Safety	None detected
General virus detection by in vitro culture	Safety	None detected
General virus detection by in vivo culture	Safety	Negative
Specific virus detection by PCR drug substance specifications	Safety	Negative
Endotoxin (limulus amebocyte lysate)	Purity	$\leq$0.5 EU/mL (target depends on patient dose)
Bioburden (culture)	Purity	$\leq$5 cfu/mL
Protein concentration (A_{280} absorbance)	Strength	Target ±5% mg/mL
Cell-based bioassay (if available)	Potency	70–130% specific activity versus reference standard
Antibody binding (ELISA)	Identity; potency	Binds to target antigen; 70–130% as potency
Cation exchange chromatography	Identity; purity	Main peak amount and profile similar to standard
Size exclusion chromatography (HPLC)	Purity	>95% monomer, report percent aggregate
SDS-PAGE (reduced)	Purity	>90 area % of HC plus LC
Isoelectric focusing	Purity	Main peaks match reference profile
pH	General	Target ±0.2 at 25 °C
Osmolality	General	Target ±20% mOsm/kg

Drug product specifications		
Test	*Category*	*Specification*
Sterility	Purity	Meets 21 CFR 610.12
Endotoxin (limulus amebocyte lysate)	Purity	$\leq$0.5 EU/mL (target depends on patient dose)
Appearance	Purity	Clarity and color similar or better than reference standard
Particulate analysis	Purity	Meets US pharmacopeia requirements
Volume per vial	Strength	Delivers not less than the target volume mL/vial
Protein concentration (A_{280} absorbance)	Strength	Target ±5% mg/mL
Antibody binding (ELISA)	Identity; potency	Binds to target antigen
Cell-based bioassay (if available)	Potency	70–130% specific activity versus reference standard
Cation exchange chromatography	Identity	Main peak amount and profile similar to standards

(continued)

Table 4.1 (continued)

Drug product specifications		
Test	Category	Specification
Size exclusion chromatography (HPLC)	Purity	>95% monomer, report aggregate
SDS-PAGE (native and reduced)	Purity	Profile similar to standard, >90% HC plus LC (reduced)
Isoelectric focusing	Purity	Main peaks match reference profile
pH	General	Target ±0.3 at 25 °C
Osmolality	General	Target ±20% mOsm/kg

Abbreviations: *CHO* Chinese Hamster Ovary, *HC* Heavy Chain, *HPLC* High-Performance Liquid Chromatography, *LC* Light Chain

Table 4.2 Characterization tests for the drug product of a monoclonal antibody prior to entry into a phase 1 clinical trial

Drug product characterization tests	
Method	Characteristic
Peptide map with mass spec	Assess degradation reactions (e.g., deamidation, clips)
N-terminal sequencing	Matches predicted sequence
Western blot	Confirm SDS-PAGE bands are antibody derived
Oligosaccharide mapping	Assess structure and heterogeneity of glycosylation
Monosaccharide content	Consistency of glycosylation

Abbreviations: *CFR* Code of Federal Regulations, *HC* Heavy Chain, *HPLC* High-Performance Liquid Chromatography, *LC* Light Chain, *US* United States

- The DP used for key preclinical testing, in particular IND-enabling toxicology testing, should ideally have the same formulation and administration route and be from the same batch as that intended for clinical use. If a different batch is used (this is common), the sponsor will need to establish that the preclinical and clinical products are comparable with bridging studies.
- Reference standards should be established to support testing, and specifications established to make sure that the DP meets regulatory expectations for identity, purity, potency, and strength (strength refers to the amount of the active ingredient).
- Additional characterization may be performed that is not required as a specification, for example, analysis of carbohydrates or protein sequencing of a DP lot.
- Formal stability testing is needed to establish that DP characteristics do not change before administration (see regulatory guidelines listed in Sect. 4.3 Resources).
- Tests to measure host cell protein and DNA impurities are needed. These tests are commercially available for common host cell lines but can be complex and expensive to develop for novel culture systems.
- Most assays do not need to be validated for phase 1 studies, but key assays should be qualified and demonstrated to be scientifically sound. Assays related to safety will receive greater scrutiny.

- After manufacturing, the QA unit reviews the manufacturing batch records and test results to assure that the drug substance and drug product are within specification and investigates any discrepancies in results or in GMP compliance. Adventitious agent testing for a biological product takes about 2 months, and full batch release typically takes 3–4 months after the drug substance is produced before the product is released for clinical use.

> **Box 4.7: What Surprised an Academic?**
> In the spirit of "No task is too small" and "All employees are equal" in the early days of KAI, I unboxed a shipment of thermometers and threw away the shipment materials and other, what I considered trivial, information about the thermometers. You could later find me with the VP of Drug Development digging through the company's dumpsters for the certificate documents that were essential to keep, as these thermometers were to be shipped to the clinical sites to ensure that the trial drug was stored at the right temperature. *—DM-R*

4.3.6   Advanced Therapy Products: Gene and Cell Therapies

In the years since the first publication of this book, the promise of genetic engineering has been fulfilled in remarkable new ways, as academic and industry scientists bring new generations of gene therapy vectors and genetically manipulated cells to patients. These advanced therapies take advantage of the power and specificity of genetic methods to provide a patient benefit, either based on vector transfection of a patient's cells after administration (a classical gene therapy approach, in which the vector is the DP and corrects a genetic deficiency) or administration of cells that have typically been modified by genetic methods ex vivo and then administered. From this perspective, most cell therapies are also gene therapies and are subject to gene therapy regulations. Unlike the antibody technology described above, CMC expectations for advanced therapies are evolving rapidly and developers may need extra communication with FDA to make sure their efforts are on the right track, especially if an accelerated approval pathway is planned. Thorough documentation is needed, and quality assessment of the product and the materials used in manufacturing will be even more critical for success.

The CMC development of a therapeutic gene therapy vector overlaps in many ways with the steps needed to develop and manufacture a prophylactic vaccine: Bioreactors are used to grow host cells that are infected with viruses or plasmids to produce active vectors, such as adeno-associated virus (AAV) or lentivirus. In addition to viral vectors, some gene therapy approaches involve direct administration of nucleic acid, for example, aided by electroporation or using lipid-based nanoparticles, as seen in mRNA-based vaccines for COVID-19. The first gene therapy vectors approved by FDA were Luxturna in 2017 and Zolgensma in 2019 (AAV-based therapies) to treat a genetic retinal disorder and spinal muscular atrophy in infants, respectively. Also in 2017, the CAR T cell therapy products Kymriah and Yescarta were approved to treat B cell lymphomas. Since the approval of these therapeutics, numerous products have entered clinical development to address other diseases.

Currently, a wide variety of novel cell therapies and cell engineering techniques have been developed; FDA reported in 2020 that about 200 new product INDs were filed to join more than 800 active INDs supporting clinical testing (Commissioner O of the 2020). A thorough review of relevant CMC topics and product types is beyond the scope of this chapter, but some topics are discussed here to reflect current approaches and regulatory considerations:

- Cell therapies are distinct from other biological products in that the manufacturing process begins with human cells; CMC considerations begin when cells are collected, with documentation of the donor's health status and maintaining records of all storage and handling.
- Much greater attention is also needed to support clinical administration; studies should demonstrate that the patient is receiving the intended dose of viable and potent cells. Cell products should have a viability specification of at least 70%.
- Because cell products cannot generally be sterilized, manufacturing must be aseptic from start to finish and developers will need to perform mock runs using culture media to demonstrate the ability to maintain sterility.
- Cell passage numbers should be tracked carefully to establish process limits in manufacturing; cells at the maximum planned passage number should be characterized to demonstrate consistency.
- FDA issued a comprehensive guidance to assist developers using gene therapy vectors or gene-modified cells to treat disease: "Chemistry, Manufacturing and Controls Information for Human Gene Therapy Investigational New Drug Applications" (January 2020). Developers are advised to review this carefully before initiating construction of vectors or cell editing. While an FDA guidance typically uses "should" rather than "must" language, developers should provide data and rationale to support their position if the topics in the guidance are not addressed as recommended.
- Cell editing and manufacturing may involve complex biological materials, for example, bacterial plasmids used to create viral vectors, guide RNAs and restriction enzymes for gene editing, and cytokines to support cell growth and differentiation. These materials are considered ancillary materials (they are not part of the products themselves) but carry risks to safety if contaminated or impure. Therefore, regulators increasingly expect developers to use components either manufactured under full GMP or produced and tested with strong controls and documentation to assure performance and safety.
- If a component like a viral vector is used to edit cells ex vivo and will be supplying a gene to the subject that persists, the vector can be considered a drug substance (active ingredient) and should be manufactured and characterized accordingly.

Cell therapies are either autologous (cell processing begins with the patient's own cells, for example a CAR T therapy) or allogeneic (the product can be used in patients other than the cell donor). The organization and challenges of autologous cell therapy manufacturing are substantially different from manufacturing allogeneic cell products or other biologicals:

- Logistics support for acquisition and transport of patient cells and delivery of the manufactured product back to the patient is critically important. Strong information systems and labeling are needed to prevent mix-ups.
- Processes and specifications may be more variable, for example, due to different ages or prior treatments in patients providing the donor cells for processing.
- To provide supplies for larger trials or commercialization, developers need to "scale out" (use many small bioreactors) rather than scaling up into larger culture vessels.
- Automation of cell culturing and processing has become critically important to achieve consistent production and to avoid open manipulations and connections that risk contamination. The use of closed systems has allowed manufacturing in facilities with less stringent environmental controls and product segregation approaches.

4.3.7 Summary

The activities listed above are expensive and somewhat daunting for a first-time developer, but new biological products regularly pass through these steps with success. Experienced CDMOs and testing labs are in place to help novel products and startup companies reach the clinic. The rewards can be spectacular: Biologicals have changed the practice of medicine and the course of many serious diseases. In 2020, six of the ten top-grossing pharmaceuticals were biologicals, with estimated worldwide sales of over $55 billion (Sagonowsky 2021). In the following year, novel mRNA vaccines, developed on a greatly accelerated timeline, delivered remarkable results in helping turn back the COVID-19 pandemic. Promising new biological products, such as antibody–drug conjugates, fusion proteins, therapeutic and oncolytic vaccines, and new gene and cell therapy applications are in

Resources
- ICH Quality Guidelines (relevant both to US and international regulatory requirements): https://www.ich.org/page/quality-guidelines.
- Points to Consider in the Manufacture and Testing of Monoclonal Antibody Products for Human Use: http://www.fda.gov/downloads/BiologicsBloodVaccines/GuidanceComplianceRegulatoryInformation/OtherRecommendationsforManufacturers/UCM153182.pdf.
- Points to Consider in the Characterization of Cell Lines Used to Produce Biologicals: http://www.fda.gov/downloads/BiologicsBloodVaccines/GuidanceComplianceRegulatoryInformation/OtherRecommendationsforManufacturers/UCM062745.pdf.
- Points to Consider in the Production and Testing of New Drugs and Biologicals Produced by Recombinant DNA Technology: http://www.fda.gov/downloads/BiologicsBloodVaccines/GuidanceComplianceRegulatoryInformation/OtherRecommendationsforManufacturers/UCM062750.pdf.

- Points to Consider in the Collection, Processing, and Testing of Ex-Vivo Activated Mononuclear Leukocytes for Administration to Humans: http://www.fda.gov/downloads/BiologicsBloodVaccines/Guidance ComplianceRegulatoryInformation/OtherRecommendationsfor Manufacturers/UCM062770.pdf.
- Supplement to the Points to Consider in the Production and Testing of New Drugs and Biologic and Produced by Recombinant DNA Technology: Nucleic Acid Characterization and Genetic Stability: http://www.fda.gov/downloads/BiologicsBloodVaccines/Guidance ComplianceRegulatoryInformation/OtherRecommendationsfor Manufacturers/UCM062777.pdf.

development and will require a similar CMC effort to establish a solid technical and regulatory foundation and fully deliver their benefits to patients around the world.

4.4   Clinical Trial Design

Lyn Frumkin and Ted McCluskey

Because clinical trials involve the safety of human participants and are expensive, it is important to plan carefully to avoid costly trial amendments (time and money), discontinuation of the trial, or results that do not clearly answer whether the treatment was safe or effective. The latter issue is especially important in resource-limited studies, that is, an investigator-sponsored trial, where a physician investigator initiates the trial and holds the IND rather than a company-sponsored trial with more financial and operational resources. In an investigator-sponsored trial, recruiting an adequate number of study participants can be challenging. However, a study that cannot assess safety and efficacy because of an inadequate number of planned study participants should not be conducted on ethical grounds. Here, we discuss issues that will support the design of proper pilot studies with small sample sizes that provide meaningful information on the safety, tolerability, and efficacy of a treatment.

4.4.1   Classical Phased Model of Drug Development

Once a new agent has been successfully tested in multiple animal models for safety and toxicity, and manufacturing has met regulatory specifications, an IND application can be filed with FDA in anticipation of the first human trial. After the application is accepted and FDA agrees that the trial may proceed, human clinical trials can begin. When a person or company submits the IND application, they become the responsible party ("sponsor"). The initial clinical trial of a new investigational compound usually begins with a phase 1 study in 40–60 healthy adults to assess safety, tolerability, and pharmacokinetics (PK) using single-ascending doses given one time and multiple-ascending doses usually given for up to 14 days. An exception is the use of a new molecular entity (NME) in patients with cancer if the toxicity of the drug does not allow it to be used in healthy volunteers; in this situation, the initial study must be done in patients. The goals of the initial study are (1) to identify adverse effects and determine the maximum tolerated dose and (2) to look for preliminary evidence of efficacy. Phase 2 studies follow to estimate initial efficacy and safety in patients and to determine best endpoints, doses, and "signal strength." Phase 3 studies confirm efficacy, usually in at least two well-conducted studies with robust and clinically meaningful results (or a single trial with more stringent requirements of efficacy if addressing a serious condition). Phase 3 studies also allow evaluation of the safety profile in a larger number of patients. The number of patients enrolled in phase 2 or 3 trials varies dramatically depending upon the estimated event rate, treatment effect, placebo response, types of controls, prevalence and seriousness of the condition under study, and other prestudy variables.

4.4.2   The SPARK Model

In SPARK, the investigator holds the IND and is the responsible party. Investigator-sponsored trials typically have fewer resources available than industry- or government-sponsored trials. These studies are often conducted at one or a few academic centers, have limited participant numbers, and seek biological signals rather than definitive clinical outcomes for the initial first-in-human trial. However, in contrast to phase 1 healthy adult studies involving an NME, these phase 1 trials (often called phase 1b) typically use drugs that are repurposed, a strategy for identifying new uses for approved (or investigational) drugs that are outside the scope of the original medical indication. This approach is advantageous in allowing an initial evaluation of efficacy in a patient population using existing knowledge (manufacturing, safety, and drug interactions) that may reduce the time and cost of clinical trials leading to approval.

4.4.3   The Clinical Protocol

The clinical trial protocol plays a key role in study planning, conduct, interpretation, oversight, and external review by detailing the plans from the initial ethics approval before the study enrolls participants to the analysis and dissemination of results (U.S. Food and Drug Administration 2018). A well-written protocol facilitates assessment of the scientific, ethical, and safety issues before a trial begins; consistency and rigor of trial conduct; and full appraisal of the conduct and results after trial completion. The importance of a well-written protocol has been emphasized by journal editors, reviewers, researchers, and public advocates.

Clinical trial protocols are read by many audiences with very different skills and concerns, including investigators and staff conducting the trial; IRB or outside the US, Ethics Committee members, who approve the trial before study start; regulatory agency physicians and scientists; public advocates; the general public (if posted on a clinical trial registry or published as an appendix in a journal with detailed study results); and industry or venture capital firms doing due diligence as part of funding or licensing evaluation. Protocol deficiencies can lead to poor trial conduct and costly protocol amendments and can jeopardize both study interpretation and subsequent publication. The most common deficiencies in protocols include inadequate descriptions of the primary outcome, the use of blinding (masking treatment assignment for participants or study staff), methods for reporting adverse events, sample size calculations, and prespecified (as opposed to after the study) data analyses (Chan et al. 2013a).

4.4.4   Where to Begin

There are many excellent clinical protocol templates available online, as well as complete protocols appearing as appendices to journal articles describing the results of late-stage trials. These complete protocols, however, may include some sections not relevant for early-stage trials. Two templates for clinical trial protocol development we recommend for SPARKees are publicly available on the Stanford SPARK website (see Sect. 4.4 Resources). These templates, with guidance language, are consistent with recommendations from both the ICH (U.S. Food and Drug Administration 2018) and the SPIRIT (Standard Protocol Items: Recommendations for Interventional Trials) initiative (Chan et al. 2013a, b). The latter was developed in consultation with key stakeholders including trial investigators, healthcare professionals, statisticians, journal editors, ethicists, and regulatory agencies to identify key deficiencies in protocol content. The SPIRIT initiative led to a guideline checklist for the minimum content of a clinical trial protocol (Chan et al. 2013a).

4.4.5   Pre-IND Meeting

A pre-IND meeting with FDA before the first human study can be valuable to verify all registration requirements and discuss safety considerations, endpoints, and safety or efficacy measures to be included in the clinical development program. FDA will typically grant the pre-IND meeting within 60 days after receiving the meeting request, and these additional 2 months should be incorporated into the development timeline. After the meeting is scheduled, FDA requires that the sponsor submit a briefing package that must arrive at least 30 days before the meeting, outlining the preclinical work and initial clinical trial designs. Pre-IND meetings can be in person, by teleconference, or more frequently, by written response only (WRO). When planning a pre-IND meeting, it is best to identify the key issues that need resolution, suggest the answer that is desired, and ask if FDA agrees (e.g.,

"The sponsor believes that three doses can safely be studied in phase 1 based on preclinical toxicology data—Does the Agency agree?"). This will prompt clear written answers from FDA to clarify development questions that the sponsor may have at this early stage.

4.4.6   Phase 0 Trial

The term "phase 0 trial" is applied to two different scenarios. In each case, conducting the phase 0 trial is intended to inform the subsequent clinical development plan before beginning standard phase 1 or phase 2 trials.

The term "phase 0 trial" is commonly applied to a clinical study conducted to observe the natural history of a disease in patients receiving current standard of care treatment. The study is often conducted to better understand the disease's impact on surrogate or clinical endpoints over a specific timeframe in the intended patient population. This information can be indispensable in informing the design of future trials. Such phase 0 trials can be particularly helpful in establishing disease variability (how the disease fluctuates over time) among patients. For instance, a phase 0 trial in psoriasis patients experiencing a disease flare might show that psoriatic plaques resolve as part of the natural history of the disease; therefore, using plaque size as an endpoint could overestimate the treatment effect in a short trial. Knowledge of disease variability and progression enables better prediction of the number of patients needed in future trials to demonstrate the desired treatment effect and how long to monitor study participants to detect this effect. Sometimes this information is available in published literature, but a rare disease (e.g., an orphan indication affecting less than 200,000 people in the US) may not have much data available. A phase 0 trial is also a good opportunity to test the measurements and tools planned for use in the clinical trial and to verify that the trial sites are adequately trained in the proper use and reporting of both.

"Phase 0 trial" can also have a specific regulatory meaning. FDA introduced the concept of phase 0 trial, also called an exploratory IND, in 2006 to evaluate subtherapeutic doses of new chemical entities in healthy volunteers or patients. Typically, a "microdose" of drug (~ 1/100th of the anticipated efficacious dose or of the No Adverse Effect Level dose in preclinical studies) is administered to a limited number of participants for no more than 7 days to characterize pharmacologic activity. Such studies may help assess (1) PK and biodistribution, often using various imaging technologies, and (2) PD (target engagement and mechanism of action). Up to five structurally related compounds can be studied simultaneously, so the results may also be used to select which agent should be advanced into a standard phase 1 study. In cancer patients, phase 0 studies may be conducted prior to tumor resection to assess whether acceptable drug concentrations are present in the resected tumor. Because an extremely low dose is only administered a single time, exploratory IND studies may be performed earlier in the course of IND-enabling studies.

4.4.7 Phase 1 Studies

Phase 1 trials are typically the first time an NME is tested in humans ("first-in-human" or FIH trial). As such, the goal of the trial is to characterize the safety and PK profile, including whether the profile seen in preclinical animals is also applicable to humans. Evaluating safety and tolerability are the primary concerns and the trial will be limited in size (maybe 40–60 healthy adult participants, including women who are not pregnant or nursing) and duration of treatment (usually cohorts of participants receiving single doses followed by subsequent cohorts receiving dosing for up to 14 days). The interaction of the drug with human metabolism is also typically explored by characterizing the PK and pharmacodynamics (PD). For oral dosing, food effects on drug absorption should also be studied. Because these initial trials are limited in scope and do not evaluate efficacy (outside the setting of cancer), they are usually conducted in specialized phase 1 trial units where healthy participants can be carefully monitored for adverse events and pharmacokinetics can be established through multiple carefully timed blood samples (or other measurements) obtained at frequent intervals.

FDA has provided a guidance document on selecting the maximum recommended starting dose for FIH studies based on the highest nontoxic animal dose (U.S. Food and Drug Administration 2005). A maximum "no observed adverse effect level" (NOAEL) dose is established for each animal species studied during IND-enabling preclinical safety studies. These NOAEL doses are converted to human equivalent doses (HED) using allometric scaling based on body surface area. The FIH starting dose is typically set at 10% of the lowest calculated HED. The dose is then escalated in successive cohorts until physiological effects occur. In a single-ascending dose (SAD) trial, participants are given a single dose and data/samples are collected typically for >5 biologic half-lives. Following the single dose, a nontreatment recovery period occurs to collect safety data. In a multiple-ascending dose trial (MAD), participants receive repeat doses and the safety and PK are observed. In both SAD and MAD, new cohort groups are dosed at higher levels until a lack of tolerability occurs, a full therapeutic effect is noted, or the maximum planned dose is reached without intolerability.

Phase 1 studies involving healthy participants are usually randomized, placebo-controlled, and double-blinded. Randomization refers to a method based on chance by which study participants are assigned to a treatment group. Placebo-controlled is when an inactive substance (placebo) is given to one group of participants, while the active drug being tested is given to another group. A double-blind study is one in which neither the participants nor study staff know which treatment the participant is receiving. These features are all intended to minimize bias and differences among groups that can limit proper data interpretation. Phase 1 studies also employ sentinel dosing, a strategy where one person receiving active drug and one person receiving placebo in the first cohort of participants is dosed in advance of the remaining participants. This ensures that if adverse events manifest quickly, as few participants as possible are impacted.

After the first phase 1 SAD and MAD trials provide an initial readout of safety and PK, additional phase 1 studies are sometimes conducted to look at drug–drug interactions, if other medications are frequently co-administered in this indication; cytochrome P450 (CYP) inhibition, if metabolism is important for drug elimination; and food interactions if the drug is administered orally. These trials look for factors that alter the investigational drug PK and may change the safety profile in a subset of patients. Knowledge gained here may lead to changes in clinical trial design or exclusion of certain participants from larger phase 2 and 3 efficacy trials.

At the end of the phase 1 trials, the investigations should have established the range of safe doses and, in some cases, a maximum tolerated dose (MTD). In nontoxic therapies, an MTD is often not determined for the highest human dose. Also, the PK (absorption, distribution, metabolism, and excretion) should be known, and an acceptable route of administration established.

Phase 1 study goals for a drug that is being repurposed for a new indication may differ. If the drug is being administered as previously approved, the safety profile is well established. But if a new formulation, higher doses, or a different route of administration are to be used for the new indication, further phase 1 studies are conducted to assess safety, tolerability, and PK of the new approach. Such studies are usually conducted in patients and include some preliminary efficacy outcomes as secondary endpoints. In contrast to the placebo-controlled nature of healthy adult SAD/MAD studies, these studies (referred to as phase 1b) can have various types of controls (e.g., none, placebo, active comparator, standard of care only). Because these studies are often limited in resources, surrogate endpoints can play an important role. Here, special attention is given to evaluate whether surrogate endpoints can be used to detect effects that may translate into more clinically relevant outcomes in later-stage trials.

4.4.8 Surrogate Endpoints

The endpoints selected to monitor safety and efficacy not only help determine the success of the trial, but also factor in the duration, planned number of participants needed, cost, and site selection for the study. The most important feature in choosing a primary endpoint is that the endpoint reflects whether the intervention provides a clinically meaningful benefit. As specified by federal law, FDA requires "substantial evidence" of effectiveness before approving a novel therapeutic. The agency accepts clinical endpoints that reflect important patient benefits (e.g., survival, function, how the patient feels) or validated surrogate endpoints that have been shown to predict clinical benefit (e.g., undetectable viral RNA levels at 48 weeks for HIV infection) as the basis for traditional approval (Human Immunodeficiency Virus-1 Infection: Developing Antiretroviral Drugs for Treatment Guidance for Industry 2015).

A surrogate endpoint is an indirect measure that is expected to reflect with high accuracy more clinically relevant endpoints (e.g., hospitalization or survival). Examples include laboratory values (change in blood glucose for diabetes mellitus), radiographic imaging (change in tumor size on magnetic resonance imaging for cancer), or clinical measures (how far one can walk in 6 min for heart failure). Surrogate endpoints that measure biological processes (such as blood glucose) are called biomarkers. (See Sect. 2.11 for more on biomarkers.)

Table 4.3 Categories of outcome measures, according to the level of evidence for efficacy

Categories of outcome measures		
	Level of evidence for efficacy	*Example*
Level 1	True efficacy measure	Death or hospitalization in pulmonary arterial hypertension or heart failure
Level 2	Validated surrogate	Six-minute walk distance in pulmonary arterial hypertension
Level 3	Nonvalidated surrogate "reasonably likely to predict clinical benefit"	Progression-free survival in some settings of cancer
Level 4	Correlate that measures biological activity but has not been established to be clinically meaningful at later stages	Negative blood cultures in treating various infectious diseases

The advantage of using a surrogate endpoint, especially in early-stage trials, is that it may allow recruitment of fewer participants because of a stronger treatment effect, less variability, and higher response rate, thus making it easier to detect a true treatment effect in a small trial of short duration. The value, however, is highly dependent upon how strongly the surrogate predicts a more clinically meaningful outcome in later-stage trials. For a validated surrogate, the strength is so high that regulators will accept the endpoint as the primary basis of approval. One example is the use of blood pressure readings for antihypertensive drugs instead of more clinically meaningful endpoints of cardiovascular morbidity or mortality.

Fleming and Powers (Fleming and Powers 2012) have provided examples of different categories of outcome measures, according to the level of evidence for efficacy. Levels 2, 3, and 4 are considered "indirect" endpoints. Some examples are included in Table 4.3.

Some surrogate endpoints—for example, blood pressure (for hypertension), cholesterol levels (for elevated cholesterol), or 6-minute walk distance (for pulmonary arterial hypertension)—are highly validated and acceptable endpoints for pivotal phase 3 drug registration trials with FDA, but others are not considered sufficient to validate efficacy and safety because they measure only a small aspect of a complex human disease. Although FDA favors hard clinical endpoints, the use of a nonvalidated surrogate endpoint can still be useful in early-stage trials to complement other endpoints if it may predict clinical outcomes.

Finally, when considering endpoints, it can be challenging to determine what a clinically meaningful effect is. For example, assume there is an ordinal pain scale measuring 1–10 (with 10 being the highest level of pain). With study treatment, how would you interpret the meaningfulness of a mean change from baseline to week 8 of 2.8? With pain outcomes, clinicians and regulators agree that a 50% reduction in pain is a clinically meaningful outcome. As a result, a "responder endpoint" would be the proportion of patients with a 50% reduction in pain at treatment completion. Similarly, for drugs intended to treat nicotine dependence, the proportion of participants who achieve continuous abstinence (confirmed by exhaled carbon monoxide) during the last 4 weeks of treatment is recognized as more clinically meaningful than mean change from baseline in the number of cigarettes smoked per week.

While differences between active drug and control groups would still need to be assessed for the minimal clinically important difference, an endpoint where patients are either "responders" or "nonresponders" based on whether they cross predefined thresholds is advantageous when available. Such outcomes build in the concept of clinical meaningfulness; the proportion of patients who are responders serves as an easy-to-interpret measurement of the treatment's effectiveness. In addition, unlike "change from baseline" endpoints, responder endpoints that involve "proportion of participants" do not require estimations of variability for sample size calculations. In small clinical trials, however, responder endpoints should be supplemented by other endpoints as they use less data than a continuous endpoint and may in some circumstances require more patients to show the existence of treatment effects.

4.4.9   Importance of Biostatisticians and Clinical Trialists

Just as in preclinical animal studies, it is important to involve a biostatistician and clinical trialist in discussing both study design and the number of study participants (sample size) to detect a true treatment effect. A sample size calculation does not indicate what the result will be, but rather is a prestudy exercise that conveys the degree of confidence that a given number of participants will be adequate to detect a predetermined treatment effect at a prespecified statistical significance level. This adequacy is determined in part by agreeing on what is usually a high probability to conclude that a treatment has efficacy when it truly does (power), usually 80% or 90%, and a low probability to falsely conclude a treatment has efficacy when it truly does not (alpha, usually 5% or less). Pilot studies with small numbers of participants may have lower power than desired (i.e., decreased ability to detect a true treatment effect) at a standard significance level.

Sample size calculation relies on subjective choice of four variables:

1. Estimated treatment effect.
2. Number of participants per group.
3. Power is the probability of detecting a given treatment effect or greater, if one truly exists ("true positive"). Low power can lead to falsely rejecting an effective therapy (Type II error).
4. Significance level, or alpha, is the probability of falsely detecting a treatment effect when none truly exists ("false positive"). This can lead to falsely accepting an ineffective therapy (Type I error).

Common misconceptions for not conducting sample size calculations:

- "Study is of a pilot nature."
- "Treatment effect of investigational drug is unknown."
- "Standard deviation for effect estimation is unknown."
- "Unable to power study to be statistically significant."
- "This is a safety study" (when it isn't).

Power is exclusively a pretrial concept to estimate the probability of predetermined results or greater under a specified hypothesis. A common error is estimating a treatment effect that is unrealistically favorable to justify both a small number of evaluable patients and a desire for 80% power. Another common error is expecting a biostatistician to determine what the estimated treatment effect is; rather, the clinical investigator or clinician must provide the effect to the biostatistician to help determine the number of participants needed to detect this effect or greater at a significance level. This treatment effect is ideally the "minimal clinically important difference" (MCID) between active and placebo groups, which is the smallest difference in an outcome measure that is clinically meaningful. If the study is powered to detect a highly robust effect, it may not have enough participants to detect a smaller but clinically relevant effect.

As an example, patients with pulmonary arterial hypertension have impairment in exercise capacity. How far one can walk in 6 min is considered a validated surrogate endpoint for more clinically relevant endpoints, such as time of clinical worsening or hospitalization. An approximately 35-m mean change from baseline to final assessment (with a standard deviation of 60 m) is considered the minimum clinically relevant difference; as a result, a change of 10 m would not be considered meaningful and a change of 50 m would be viewed as highly robust. Studies using exercise capacity (6-min walk distance) in pulmonary arterial hypertension are designed with reasonable power (80–90%) to detect this MCID or greater at a significance level if it truly exists. Having too few participants to detect the MCID is called an "underpowered" study; it does not have the optimal ability to detect, at a significance level, a true treatment effect that is clinically meaningful.

Because academic trials are often resource-limited with fewer numbers of participants, such trials may not be able to detect a true and clinically relevant treatment effect with high probability (underpowered). At a given sample size, factors that contribute to lowered power to detect a given clinically relevant effect include high placebo response rates, low event rates, modest treatment effects, and benefit from slowing of worsening (as opposed to improvement per se). Despite this concern, academic studies can have sufficient ability to evaluate for preliminary evidence of safety, efficacy, and variability, including the presence of a biologic signal.

Box 4.8: Common Pitfalls to Avoid
1. Not appreciating the value of sample size determination.
2. Not incorporating surrogate endpoints, including biomarkers, that may help determine a biologic effect in an early-stage trial.
3. Starting with too high a dose or escalating the dose too quickly in early safety studies.
4. Trying to make definitive efficacy conclusions with an underpowered study.
5. Working "backwards" with a given number of study participants in sample size calculations to have high power to detect an unrealistically high treatment effect.
6. Retroactively looking through subsets of data to try to show efficacy.
7. Trial sites that deviate from the protocol in enrollment criteria, record keeping, or measurements.
8. Poor patient retention and missed follow-up visits.

4.4.10 Phase 2 Studies

The goal of the phase 2 trial(s) is to obtain preliminary but more definitive data than from phase 1b trials regarding whether the intervention is safe and effective in the target patient population. This includes studying the conditions (final dose, route, and regimen) and the endpoints (primary, secondary, and surrogate) that will be used in phase 3 confirmatory trials, as well as continuing to build the drug safety profile.

Because phase 1 trials have established some level of safety and tolerability (and possible efficacy in a phase 1b trial of a repurposed drug), phase 2 studies will have larger participant numbers and typically longer duration. Phase 2 trials usually study a range of doses in parallel. The dose range is highly dependent on the safety and PK results obtained in phase 1. Often, surrogate endpoints will be used for evidence of efficacy since they typically require both a smaller sample size and a shorter trial duration than clinical outcomes. A distinction is sometimes drawn between a phase 2a trial that studies dose and regimen selection to determine the maximum dose and a phase 2b trial that focuses on measures of efficacy to find the minimally effective dose.

As with phase 1b trials, options for control groups in phase 2 trials include:

- No control group (open label = unblinded).
- Placebo group (open label or blinded).
- Active treatment with a specific (usually approved) drug.
- Standard of care (optimal treatment available).
- Standard of care (clinical practice in real-world setting).
- A crossover design, advantageous for rare diseases or where recruitment is changing, in which each participant serves as their own control by receiving each treatment (such as placebo followed by active drug) in succession separated by a brief period without study drug. An advantage of this internal control (i.e., within each participant) is eliminating individual participant differences from the overall treatment effect, thus enhancing statistical power with fewer participants in a rare condition. However, in a crossover study, it is important that the underlying disease does not significantly change over the time studied as part of the natural history and that the effects of one treatment are gone before the next is applied.

Box 4.9: What Surprised an Academic?
When planning our multicenter phase 2 clinical trial targeting patients with ST elevation myocardial infarction (STEMI), we asked our investigators and study coordinators at each participating hospital for a realistic prediction of how many patients they could enroll per month. Their answer was invariably four to five-fold higher than their actual enrollment rate. Overconfident study staff failed to account for a number of factors that limited patient recruitment, such as the challenges of enrolling patients in the middle of the night when study staff were not available, other trials competing for the same patients, the

fact that only 25% of STEMI patients would meet the eligibility criteria, etc. When identifying clinical sites and planning for patient recruitment, it is essential to be very detailed in determining the realistic rate of patient enrollment. Even then, discount the predicted enrollment rate by 50–75%. An overly long period of patient recruitment will greatly increase the cost of your clinical trial, and may even lead to failure of your study if you run out of resources. —*KVG*

As previously noted, blinding both the research team and patients to the treatment assignment is very important in clinical trial design: not only to prevent investigator or study staff (unintentional) bias influencing data collection, but also to best measure any placebo effects that participants experience. In rare situations (e.g., hemodynamic indices in pulmonary arterial hypertension), there is no placebo effect. That is, patients receiving placebo historically show no change or worsened hemodynamic effects as part of the natural history of their disease. In this setting, even though such patients are enrolled on maximum stable (but suboptimal) background therapy, removing the placebo group from the study design at this stage can facilitate enrollment in a rare condition while still allowing benefit with active treatment to reflect a true improvement. A disadvantage of this approach is that the placebo-adjusted incidence of adverse events cannot be assessed in the absence of this control group.

While phase 2 studies can be single blinded (study participant does not know whether receiving active or control), double blinded (participant and trial staff do not know), or triple blinded (participant, staff, trial sponsor, and core lab analyzing specimens do not know), an unblinded Data Safety Monitoring Committee can help inform investigators how to respond to adverse events that arise during the trial.

The sample size should be determined by the number of participants needed to adequately "power" the trial to observe a clinically meaningful change at a (statistical) significance level for the main endpoint of interest. Excellent sample size calculators are available online (see Sect. 4.4 Resources), but it is strongly advised to consult an experienced clinical trial biostatistician or clinical trialist. The biostatistician will ask the clinician: "What is the smallest change in a trial endpoint that you want to be able to detect that is clinically meaningful?" From this information and published values of the endpoint in the target population, the biostatistician should be able to provide the number of participants needed to detect the desired effect. When estimating the expected value for the primary endpoint in the placebo group based on historical data, it is important to correct for recent improvements in standard of care that may not be reflected in published data from previous studies.

Multiple surrogate endpoints may be evaluated in phase 2 trials to better characterize the safety, tolerability, and efficacy. Often a surrogate endpoint is chosen as the primary endpoint in phase 2 (and the trial is "powered" on this endpoint) and the planned phase 3 clinical endpoint is evaluated as a secondary endpoint (since it is likely to be "underpowered" with the smaller size of the phase 2 trial). A common example is the use of hemodynamics as the main endpoint in phase 2 pulmonary arterial hypertension trials, requiring about 20–25 patients/group to be able to detect the minimally important clinical difference or greater at a significance level. A secondary endpoint might include the more clinically relevant endpoint of exercise capacity, which in a phase 3 trial requires about 100 patients/group. However, trends, or in some cases a very robust effect for an underpowered endpoint, can occur and provide evidence of biological plausibility or a meaningful treatment effect.

The methods to assess the clinical outcome in phase 3 should be clearly established and rehearsed at each trial site during phase 2, so that inconsistencies between sites are minimized and do not affect data integrity. A successful phase 2 program will help minimize "surprises" in clinical operations and trial conduct that may occur during the phase 3 program. Phase 2 studies are where most programs falter; in an evaluation of 9704 programs across the US and other countries from 2011 to 2020, only 28.9% progressed from phase 2 to phase 3 (Fig. 4.4) (Biotechnology Innovation Organization 2021). Some trials fail because the drug does not work, others because suboptimal study design and conduct may preclude detection of a true effect. Perhaps most fail the "wallet" test: The clinical effect is not sufficient to justify further development.

Recommendations

- As you transition from animal to human studies, be aware of the use of concomitant medications. These can lead to drug–drug interactions, which may increase or decrease your therapeutic effect or even have a direct impact on your chosen endpoints.
- If possible, try to determine the MCID as the estimated treatment effect to power your study. This way you will be able to detect a true treatment effect that is clinically meaningful at a significance level. In some cases, the number of participants needed to achieve this is not realistic in an academic or early-stage trial. In those situations, identifying clinically important effects at study completion can be helped prior to the study by:

  - Incorporating multiple biologically related endpoints.
  - Use of biomarkers that are biologically plausible.
  - Use of validated or reasonably likely validated surrogate endpoints that may be highly affected by treatment.
  - Using relevant assessments with low placebo response rates.
  - Specifying outcomes that occur more frequently.
  - Enrolling patients at higher risk.
  - Making full use of longitudinal data.
  - Prolonging nontreatment follow-up.

- Picking the wrong clinical or surrogate endpoints can obscure a true therapeutic effect or drastically increase the time and cost of running a clinical trial. Consult FDA guidance and past clinical trials in your indication (or a related indication) to see what metrics others have used.

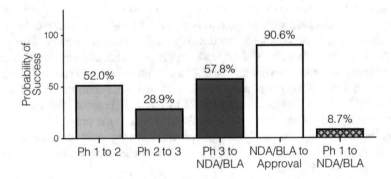

Fig. 4.4 Overall phase transition success rates from phase 1 to NDA/BLA for 9704 international programs from 2011 to 2020, adapted from Biomedtracker® and Pharmapremia®, 2020 (Biotechnology Innovation Organization 2021). Abbreviations: *BLA* Biologics License Application, *NDA* New Drug Application; *Ph* Phase

4.5   Phase 3 Studies

Lyn Frumkin and Ted McCluskey

Assuming that the phase 1 and 2 clinical trials provide reasonable evidence of safety and efficacy based on surrogate or more clinically meaningful endpoints, it is time to request an end-of-phase 2 meeting with FDA. FDA guidance will be critically important, not only for finalizing plans for the pivotal phase 3 trial(s), but also to ensure that the sponsor addresses any nonclinical concerns that must be included in the NDA.

Before registering a new drug, FDA requires two large and robust phase 3 trials to show safety and efficacy via an approved clinical endpoint with high statistical confidence. In rare cases, after discussion with FDA, highly robust treatments for serious medical conditions with great unmet needs are approved after a single phase 3 trial with more stringent safety and efficacy requirements.

4.5.1   Trial Design Trends for Small Clinical Trials

Drug development is difficult. From 2011 to 2020, only around 9% of compounds (either drugs or biologics) entering phase 1 clinical trials from FDA registration-enabling development programs successfully progressed through all clinical trial phases to submit a successful application to be registered as a new drug or biologic (Biotechnology Innovation Organization 2021). This represents a huge financial loss in research and development expenses. New trial strategies are emerging to improve measurements of efficacy and to "fail fast," that is, rapidly reach go/no-go evaluation points for safety and efficacy. This also includes novel considerations for small clinical trials that are unable to recruit large numbers of participants. The US National Academy of Medicine has emphasized the importance of balancing the need for clinical trials that are adequately powered for scientific and ethical validity with the reality that, in certain situations (rare diseases, unique study populations such as astronauts, or emergency situations), the number of potential study participants may be quite limited. Such challenges can affect study interpretation (Evans Jr and Ilstad 2001).

Small clinical trials, whether due to a low disease prevalence or limited resources, may be able to detect only gross effects and run the risk of not identifying true effectiveness and adverse events. Approaches, such as full use of longitudinal data, nontreatment follow-up, including multiple endpoints (appreciating that many comparisons increase the likelihood that a chance association could be deemed causal), and use of composite endpoints (combining several outcomes into a single outcome measure), may improve the ability to detect a true treatment effect in a small clinical trial (Day et al. 2018).

Besides the crossover design previously mentioned, there are other designs to optimize sample size. An adaptive trial design allows changes to a trial after its initiation without affecting data validity or integrity (Bhatt and Mehta 2016). Adaptive designs can be applied to all trial phases, from early-phase dose escalation to late-stage confirmatory trials. For example, in one adaptive trial design, blinded trial data by masked treatment assignment can be periodically evaluated for safety and efficacy metrics while the trial is ongoing to determine whether a sample size re-adjustment is necessary. Besides adjustments to enrollment size, premature closure of the trial for early "success" or "failure" can also be determined in some cases.

Whereas placebo is traditionally the most common type of control for a clinical trial, the use of active controls is growing and will likely increase more in the future. In clinical trials, regulations require new agents to beat placebo, not approved drugs. However, in indications where there is an approved drug on the market, head-to-head comparison with the investigational drug in a trial that requires more participants to be adequately powered can allow a direct comparison of therapeutic benefits. This benefits patients and physicians by evaluating whether a newly approved drug provides a benefit over older (possibly generic) drugs. Companies were previously reticent to run active control trials for fear their investigational drug would not be more effective than previously approved drugs. Recent legislation in the US aimed at "comparative effectiveness" will likely increase the use of active controls.

Enrollment strategies are also important. A sponsor can be a "lumper" with broad inclusion criteria for enrollment or a "splitter" that narrowly defines participant enrollment. Lumpers are often focused on maximizing potential market size after drug approval or speeding enrollment rates. Splitters, on the other hand, are worried about variation within the target indication and select for those patients most likely to benefit from the investigational drug. This latter approach is critically important in some conditions (e.g., pulmonary arterial hypertension) where certain subtypes (idiopathic, heritable, connective tissue disease, and congenital heart disease) may respond to various treatments better than other subtypes. In this case, enrolling those most likely to respond in a clinical trial can better detect a true treatment effect that might be missed by enrolling both historical "good responders" and those who poorly respond. With the demonstration of safety and efficacy in the former group, the latter group can subsequently be evaluated in separate trials. Both enrollment strategies have pros and cons; specific indications may favor one strategy over the other.

4.5.2 Final Points

Creating the proper database to collect study information and formulate plans for data analyses, data monitoring, and review of adverse events is critical to address before study start. Regardless of whether a clinical trial is investigator-, government-, or company-sponsored, it must be conducted under a protocol; the ethical principles stated in the 2013 version of the Declaration of Helsinki; applicable guidelines on Good Clinical Practice; and all applicable federal, state, and local laws, rules, and regulations (U.S. Food and Drug Administration 2018) to maximize clinical trial quality and participant safety. In addition, several important "internal" financial and operational issues must be addressed when transitioning into clinical studies. Trial costs must be estimated, and internal budget and personnel resources must be planned and accounted for to complete the initial study, including nontreatment follow-up. Considerations for patient enrollment rates at each clinical site can help predict the time required to complete the clinical trial. Insurance, IRB fees, and funds for investigator training and auditing must also be in place.

The Bottom Line

The successful execution of a clinical trial plan requires a team effort and detailed planning. Since it is the most expensive part of drug development, setbacks in execution of clinical trials (e.g., too slow enrollment, shortage of drug, delay in drug arrival to the clinical sites, inappropriate blinding, improperly designed trial, inadequate sample size) can "sink" the program for budgetary reasons, poor enrollment, or a "negative" result from the inability to detect a true treatment effect rather than because the drug is not effective.

Key Terms and Abbreviations

Key Terms

Active Pharmaceutical Ingredient (API): The pharmacologically active molecule.

Biologics License Application (BLA): Request for permission to introduce a biologic product into interstate commerce in the US.

Biosimilar: Generic version of a previously approved biologic therapeutic.

Chemistry, Manufacturing, and Controls (CMC): Manufacturing procedures and specifications that must be followed to ensure product safety and consistency between batches.

Clinical Endpoint: A measure of something a patient would experience or report.

Common Technical Document (CTD): Standardized five-module format for submission of regulatory applications to CDER and CBER.

Contract Development and Manufacturing Organization (CDMO): An external company that does both formulation development and manufacturing of a drug product.

Contract Manufacturing Organization (CMO): An external company that manufactures a preformulated drug.

Double-Blinded: Neither the participants nor study staff knows which treatment the participant is assigned to.

Drug Product (DP): Encompasses API and inactive components such as binders, capsule, *etc.*, that composes the final drug formulation of the pharmacologically active molecule administered to patients.

Drug Substance (DS): The pharmacologically active biological material—same as API and used interchangeably.

FDA Center for Biologics Evaluation and Research (CBER): Responsible for regulation of biological and related products such as blood, vaccines, and cellular and gene therapies.

FDA Center for Devices and Radiological Health (CDRH): Responsible for ensuring the safety and effectiveness of medical devices and the safety of radiation-emitting products.

FDA Center for Drug Evaluation and Research (CDER): Responsible for regulating the safe and effective use of over-the-counter and prescription drugs, including monoclonal antibodies, immunomodulators, growth factors, and other protein therapeutics.

First in Human (FIH): A clinical trial in which a treatment is tested in humans for the first time.

Good Manufacturing Practice (GMP): Exacting procedures and documentation of quality assurance carried out at a certified facility (sometimes referred to as "cGMP" for "current" practice).

Heavy Chain (HC): The large polypeptide subunit of an antibody.

Human Equivalent Dose (HED): A dose that is predicted to provide the same level of drug exposure in humans as a dose administered in a preclinical animal model provides in that species (*e.g.*, NOAEL or nontoxic dose); calculated using a species-specific allometric scale based on body surface area.

Institutional Review Board (IRB): The entity designated to review and monitor research involving human participants in the US; often called the Ethics Committee outside of the US.

International Conference on Harmonisation (ICH): Organization that provides international guidelines for drug testing.

Investigational New Drug Application (IND): Request from a clinical sponsor to obtain authorization from FDA to administer an investigational drug or biological product to human subjects.

Light Chain (LC): The small polypeptide unit of an antibody.

Master Batch Record (MBR): A written procedure of all manufacturing and testing methods that is required to be approved by QA.

Master Cell Bank (MCB): Repository of frozen cell aliquots for the clonal hybridoma line selected for biological product manufacturing.

Master File: FDA file certifying a manufacturing company based on prior IND submissions establishing GMP requirements.

Maximum Tolerated Dose (MTD): The highest dose of a drug or therapy that does not cause unacceptable side effects or toxicity.

Media Fill: Manufacturing run using media performed to demonstrate sterile operations.

Minimal Clinically Important Difference (MCID): The smallest improvement in treatment outcome that a patient would consider worthwhile.

Multiple-Ascending Dose (MAD): A trial where subjects in each cohort receive multiple administrations of the same dose of study drug, often over 14 days; the dose for each successive cohort is escalated once safety is established for the prior cohort.

New Drug Application (NDA): Mechanism to obtain FDA approval for the sales and marketing of a new drug in the US.

New Molecular Entity (NME): A drug product that is a new chemical entity not previously approved by FDA.

No Observed Adverse Effect Level (NOAEL): The maximum dose administered in preclinical safety studies where no undesirable side effects are seen; sometimes called the nontoxic dose.

Orphan Indication: An FDA designation of a disease or condition that affects less than 200,000 people in the US or for a treatment that is not expected to recoup its research and development costs due to pricing constraints.

Pharmacokinetics (PK): Measurements of what the body does to a drug (absorption, distribution, metabolism, and excretion).

Placebo: A pharmacologically inactive substance used as a comparator in some clinical trials.

Prescription Drug User Fee Act (PDUFA): US legislation that allows FDA to collect fees from drug manufacturers to fund the new drug approval process.

Quality Assurance (QA): Process to comprehensively ensure the production of a safe and effective drug product by proactively optimizing the drug manufacturing and packaging process to minimize and eliminate any defects before they can occur.

Quality Control (QC): Functional unit with the responsibility and authority to ensure that the drug product meets all pre-established quality standards before it is released; this process includes reviewing and approving/rejecting all procedures and components involved in the manufacturing process, from acquisition of raw materials to final packaging.

Randomization: A method based on chance by which study participants are assigned to a treatment group.

Recombinant DNA (rDNA): DNA constructs that have been artificially manipulated and do not naturally occur in organisms.

Single-Ascending Dose (SAD): A trial where subjects in each cohort receive a single administration of a given dose of study drug; the dose for each successive cohort is escalated once safety is established for the prior cohort.

Sponsor or Applicant: Individual or entity that submits an IND, NDA, or BLA.

Stability Indicating Assay: Product testing that is performed at designated time intervals to ensure that API remains intact within prespecified levels and that impurities remain below prespecified levels. Often done at both the recommended storage temperature and a higher temperature (*e.g.*, 37 °C), which may accelerate degradation of the drug product.

Surrogate Endpoint: An indirect measure that may be a lab value, imaging, or clinical measure that is expected to reflect more clinically relevant endpoints.

Therapeutic BLA: Application submitted to CDER for products such as monoclonal antibodies, cytokines, and growth factors.

US Food and Drug Administration (FDA): The federal Health and Human Services agency responsible for protecting public health.

Validation (or Qualification of) Assays: A formal process demonstrating that an assay is specific, reproducible, and precise.

Working Cell Bank (WCB): Frozen aliquots of cells for active use in manufacturing and testing.

Key Abbreviations

AAV	Adeno-Associated Virus
CHO	Chinese Hamster Ovarian
CFR	Code of Federal Regulations
CAPA	Corrective and Preventive Actions
CYP	Cytochrome P450
DF	Diafiltration
EMA	European Medicines Agency
EU	European Union
EBR	Executed Batch Records
HPLC	High-Performance Liquid Chromatography
NIST	National Institute of Standards and Technology
PMDA	Pharmaceuticals and Medical Devices Agency
PAI	Pre-Approval Inspection
PD	Process Development
QP	Qualified Person
QSR	Quality System Regulations
RFP	Request for Proposal
SOP	Standard Operating Procedures
SPIRIT	Standard Protocol Items: Recommendations for Interventional Trials
TSE	Transmissible Spongiform Encephalopathy
UF	Ultrafiltration
WRO	Written Response Only

References

Bhatt DL, Mehta C (2016) Adaptive designs for clinical trials. N Engl J Med 375:65–74

Biotechnology Innovation Organization (2021) Clinical development success rates and contributing factors, 2011–2020. Washington DC. Available at: https://go.bio.org/rs/490-EHZ-999/images/ClinicalDevelopmentSuccessRates2011_2020.pdf

Chan A-W, Tetzlaff JM, Altman DG et al (2013a) SPIRIT 2013 statement: defining standard protocol items for clinical trials. Ann Intern Med 158:200–207

Chan A-W, Tetzlaff JM, Gøtzsche PC et al (2013b) SPIRIT 2013 explanation and elaboration: guidance for protocols of clinical trials. BMJ 346:e7586

Cohen SN, Chang ACY, Boyer HW, Helling RB (1973) Construction of biologically functional bacterial plasmids in vitro. Proc Natl Acad Sci 70:3240–3244

Commissioner O of the (2020) Statement from FDA Commissioner Scott Gottlieb, M.D. and Peter Marks, M.D., Ph.D., Director of the Center for Biologics Evaluation and Research on new policies to advance development of safe and effective cell and gene therapies. In: FDA. https://www.fda.gov/news-events/press-announcements/statement-fda-commissioner-scott-gottlieb-md-and-peter-marks-md-phd-director-center-biologics. Accessed 28 July 2022

Day S, Jonker AH, Lau LPL, Hilgers R-D, Irony I, Larsson K, Roes KC, Stallard N (2018) Recommendations for the design of small population clinical trials. Orphanet J Rare Dis 13:195

Demonstrating Substantial Evidence of Effectiveness for Human Drug and Biological Products. https://www.fda.gov/media/133660/download

Evans CH Jr, Ilstad ST (eds) (2001) Small clinical trials: issues and challenges. Institute of Medicine, U.S. National Academy of Sciences. National Academies Press, Washington, DC

Fleming TR, Powers JH (2012) Biomarkers and surrogate endpoints in clinical trials. Stat Med 31:2973–2984

(2015) Human Immunodeficiency Virus-1 Infection: Developing Antiretroviral Drugs for Treatment Guidance for Industry. https://www.fda.gov/files/drugs/published/Human-Immunodeficiency-Virus-1-Infection%2D%2DDeveloping-Antiretroviral-Drugs-for-Treatment.pdf

ICH Official web site: ICH. In: CTD. https://www.ich.org/page/ctd

Khan S, Ullah MW, Siddique R, Nabi G, Manan S, Yousaf M, Hou H (2016) Role of recombinant DNA technology to improve life. Int J Genomics 2016:e2405954

Köhler G, Milstein C (1975) Continuous cultures of fused cells secreting antibody of predefined specificity. Nature 256:495–497

Sagonowsky E (2021) The top 20 drugs by worldwide sales in 2020. In: Fierce Pharma. https://www.fiercepharma.com/special-report/top-20-drugs-by-2020-sales. Accessed 29 Mar 2022

U.S. Food and Drug Administration (2005) Guidance for Industry. Estimating the maximum safe starting dose in initial clinical trials for therapeutics in adult healthy volunteers. US. FDA, Silver Spring. Available at: https://www.fda.gov/media/72309/download

U.S. Food and Drug Administration (2018) Guidance for Industry. E6(R2) Good Clinical Practice: Integrated Addendum to ICH E6 (R1). US. FDA, Silver Spring. Available at: https://www.fda.gov/media/93884/download

U.S. Food and Drug Administration (FDA) (2004) Innovation or stagnation: challenge and opportunity on the critical path to new medical products. U.S. Food and Drug Administration (FDA)

Technology Transfer and Commercialization

5

Daria Mochly-Rosen, Kevin Grimes, Judy Mohr,
Karin Immergluck, Emily Egeler, Jennifer Swanton Brown,
Nicholas Gaich, Eugenio L. de Hostos, Grace Hancock,
Mary Wang, Robert F. Booth, Julie Papanek Grant,
Leon Chen, Nina Kjellson, Haim Zaltzman, J. Jekkie Kim,
John Walker, Alan Mendelson, Peter Boyd,
and Christopher M. Reilly

From Academia to a Start-Up

| Intellectual Property & Licensing | Academic Compliance | Not-for-Profit Drug Development | Entrepreneurship & Securing Funding | Starting a Start-Up |

Daria Mochly-Rosen, Ed, Kevin Grimes, Ed.

D. Mochly-Rosen (✉) · K. Grimes
Chemical and Systems Biology, Stanford University School of Medicine, Stanford, CA, US
e-mail: sparkmed@stanford.edu; kgrimes@stanford.edu

J. Mohr
McDermott Will & Emery LLP, Menlo Park, CA, US

K. Immergluck
Stanford University, Stanford, CA, US

E. Egeler · J. S. Brown
Stanford University School of Medicine, Stanford, CA, US

N. Gaich
Nick Gaich and Associates, Morgan Hill, CA, US

E. L. de Hostos · G. Hancock · M. Wang
Calibr, Scripps Research Institute, La Jolla, CA, US

R. F. Booth
Curasen Therapeutics, San Carlos, CA, US

As academic researchers, some of us may be interested in forming startup companies to develop and market drugs or diagnostics based on our research. Others may be happy to license their technology and findings to an existing company. In either case, it is essential that your inventions are protected by filing a patent. Unless there is patent protection to provide a period of exclusive marketing after regulatory approval, a commercial entity will not be able to realize a return on the considerable investment in finances and time required for product development.

The invention must be considered novel and non-obvious at the time of patent filing. Once an invention is in the public domain (e.g., journal article, abstract, conference presentation), it is no longer considered novel by the patent granting agencies. Therefore, if you believe that your discovery has a commercial application, do not disclose information about it in public without first filing a patent. Premature public disclosure is perhaps the chief reason that promising academic inventions are deemed not patentable and therefore not developed.

Since sharing our discoveries is crucial to the advancement of science and intrinsic to the academic mission, expeditious patent filing is critical. Your institution's office of technology licensing can quickly file a provisional patent that includes your novel claims along with an abstract of your planned presentation or manuscript to be submitted for publication. Familiarizing yourself with the people and processes of your institution's technology transfer office can save you much time.

J. P. Grant · N. Kjellson
Canaan Partners, Menlo Park, CA, US

L. Chen
The Column Group LLC, San Francisco, CA, US

H. Zaltzman · J. J. Kim
Latham & Watkins LLP, Menlo Park, CA, US

J. Walker
SPARK at Stanford Advisor, Stanford, CA, US

A. Mendelson (Deceased)

P. Boyd · C. M. Reilly
Former SPARK at Stanford Advisor, Stanford CA, US

The next step is finding a commercial partner—a venture capital firm or company that will be interested in funding the development of your invention. Knowing how to evaluate potential diagnostic or drug markets and effectively pitch to investors or commercial partners is very different skill than giving a good scientific lecture. With so much intellectual property (IP) generated in academia, a well-researched commercial assessment and an organized and effective pitch for the commercial audience can make potential licensees or investors take notice. The goal is not to recapitulate the sophisticated marketing plans drawn up in industry, but to highlight the project's potential and show some industry savvy.

As you begin exploring the commercial value of your invention, it is essential that you recognize and avoid any potential conflicts of interest. Your institutional compliance office can provide guidance regarding how to comply with both institutional rules and government laws and regulations. Even if you participate in annual compliance training, make sure to refresh your memory before proceeding with commercial or clinical activities, so as to avoid the often serious consequences of non-compliance.

This chapter covers important topics that are rarely taught to academic scientists, including patent law, technology transfer, avoiding conflicts of interest, evaluating the commercial potential of your project, pitching to potential licensees, approaching venture capitalists, and funding the development of not-for-profit programs. It also discusses legal and practical considerations for academic founders of a startup company.

5.1 Intellectual Property

Judy Mohr

Like them or not, patents are a necessary part of the pharmaceutical business. Because of the high development costs incurred in obtaining regulatory approval to market a pharmaceutical product, potential corporate partners and investors insist that the product be protected by one or more patents with sufficient patent term remaining after product launch to have an exclusive market position to recoup the development costs. Furthermore, the generic drug industry is sophisticated and is rapidly able to manufacture a bioequivalent of a branded product and, possibly, a bioequivalent product with a minor modification in a non-essential ingredient to avoid patent claims covering the product. Thus, a company developing a pharmaceutical product—whether as a new molecular entity (NME), a new delivery vehicle/platform for a previously approved drug, or a new method of treatment using an already approved compound—must define a patent strategy that supports its own business objectives, yet is also mindful of the inevitable generic competition to the pharmaceutical product. This section touches on three important aspects in building any patent portfolio: the consequences of publicly disclosing the invention before filing for patent protection, patent searching to assess patentability of an invention, and freedom to operate.

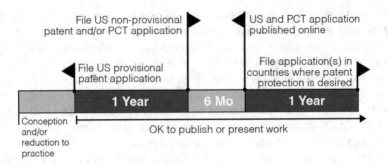

Fig. 5.1 Patent filing timeline. Abbreviations: *Mo* Month, *PCT* Patent Cooperation Treaty, *US* United States

5.1.1 Public Disclosure and Patent Filing

For an invention to be patentable, it must be "novel and non-obvious" as defined in patent law. This novelty requirement states that an invention cannot be patented if the invention is already known anywhere in the world before the filing date of a patent application for the invention. "Already known" encompasses publications, public use or sale, and oral disclosure, such as at a conference or meeting. It also encompasses an inventor's own disclosure, such as in a journal or at a conference, of the invention. Thus, it is very important to not disclose an invention in any public setting before a patent application has been filed.

Figure 5.1 depicts the sequence in a patent filing process. After an invention is made, a patent application is filed. In the US and many other countries, a provisional patent application is filed to secure a patent filing date. Once the provisional patent application is filed, the invention can be disclosed, for example, in a journal article or at a conference. One year after filing the provisional patent application, an international patent application with the Patent Cooperation Treaty (PCT) is filed, and in many cases, a US non-provisional patent application is also filed. The PCT application and US non-provisional patent applications will be published 18 months from the filing date of the provisional patent application. The PCT application is then filed in the PCT member countries where patent protection is desired.

Patent rules in the US provide for a "grace period" to file a patent application after a public disclosure of the invention by the inventor; the grace period may not apply in every situation. Also, most other countries do not grant any grace period between public disclosure and patent filing. Therefore, it is best practice to file a patent application before any public disclosure of the invention, particularly for pharmaceuticals for which worldwide market potential is desirable.

Box 5.1: What Surprised an Academic?

If you want to be sure that your invention will never benefit patients, publish your idea and data before filing a patent. Exclusive and preferably worldwide IP rights are essential to recoup the huge development costs of new drugs and biologics; without IP protection, no one will develop your drug.—*DM-R*

5.1.2 Patent Searching to Assess Patentability

A first step in determining a patent strategy for any invention is to know the "lay of the land" of publications related to the invention. This is a critical step in determining whether your invention is novel and non-obvious—two essential requirements for patentability. Your publication search should include both scientific literature and patent literature. Before beginning a search, make a list of the components of the invention and draft the proposed product label regarding how a product based on the invention will be used (i.e., the disease or condition for which FDA approval of the product will be sought and how it will be dosed). This information should track closely, if not identically, with the contents of any drafted or filed regulatory documents such as an IND application. Knowing the product components and the intended label use identifies some key words for an initial search of patent literature (issued patents and published applications) and scientific literature to obtain a solid understanding of prior publications related to the invention and thus products based on the invention.

Unless you have some familiarity with patents or patent searching, it may be easier to start searching in the scientific literature. Search terms identified using the approach above can be utilized in online databases such as PubMed and Google® Scholar. From the scientific literature search, the authors and/or institutions (academic or commercial) of the most relevant articles should be added to the list of search terms for searching patent databases.

A search of issued patents and published patent applications is best done using a combination of several databases. Patents are jurisdictional, so it is necessary to search both US and international databases to identify patent publications that may be relevant to patentability. Examination of the following three databases will give a fairly thorough search (see Sect. 5.1 Resources for website URLs).

1. **US Patent and Trademark Office (USPTO)**: The USPTO website has a Patent Public Search tool with a flexible search interface to search issued US patents and published US patent applications. The search results are hyperlinked to relevant documents, making it easy to review the documents, including highlights of where the search terms appear in the document. Granted patents and published applications can be viewed and downloaded in a familiar two column format from the Patent Public Search Tool or from other websites, such as pat2pdf.

2. **Espacenet**: This database permits searching of worldwide patents and applications, and using the "advanced search" tab, it can be searched by title, inventor, applicant (company name), etc. This database also has a tool, INPADOC (international patent documents), that will identify all counterpart patents and applications, or "family members," of a relevant patent or application. For example, if your search identifies a US patent of interest, enter the US patent number in the "publication number" search field on the advanced search page, click on the result, then look for the "INPADOC patent family" link, and click on it to see whether the US patent has counterpart filings outside the US.

3. **World Intellectual Property Organization (WIPO) PatentScope**: This database is maintained by WIPO and allows a search of PCT (Patent Cooperation Treaty) applications for international filings. Both simple and "structured" (advanced) search options are available for more than a million international patent applications. The results from a search of the databases can then be reviewed to identify documents that describe the invention in whole or in part.

5.1.3 Patent Searching to Assess Freedom to Operate

Freedom to operate (FTO) refers to whether a particular action, such as manufacturing or selling a product, can be done without infringing the valid granted patent rights of another party. Since patent rights are jurisdictional, an FTO analysis needs to be done in each particular country where the product is to be manufactured or sold.

The USPTO and Espacenet patent search sites noted above are equally useful for searching for patents potentially relevant to FTO. A key difference between assessing patentability and assessing FTO is that the latter needs only focus on the claims of granted patents, whereas the former must consider the disclosure of the entire patent document.

If the search identifies a claim in a granted patent that may be relevant to FTO, further investigation is needed to ascertain whether it poses an actual barrier to commercializing the product. For example, words in a patent claim may be subject to definitions in the patent specification (the legal term for the text in a patent that precedes the patent's claims and alters the scope of the patent claim); or admissions may have been made by the patentee while the patent application was being examined that narrow the meaning of claim terms. Also, fees are required to maintain patents in force, and if the patent owner has not timely paid the fees, the patent may no longer be enforceable. A patent attorney is typically required to advise regarding the legal scope of patent claims.

Resources
- US Patent and Trademark Office (USPTO): http://www.uspto.gov
- Pat2pdf – A FREE patent search tool: https://www.pat2pdf.org/
- Espacenet: http://worldwide.espacenet.com
- World Intellectual Property Organization (WIPO) PatentScope: http://www. wipo.int/patentscope/search/en/search.jsf

5.2 Working with the University Technology Transfer Office

Karin Immergluck

5.2.1 The Role of the Technology Transfer Office (TTO)

In the broadest sense, technology transfer in the academic context is the process of transferring knowledge, discoveries, technologies, inventions, and authored works to others, generally for the purpose of enabling further research and development of new commercial products and services. This is accomplished by various means, including publication of research results, presentations and discussions at conferences, and exchange of students or personnel between academia and industry. From the perspective of the university technology transfer office (TTO), however, the focus is on the formal transfer of IP and other technology rights, usually to a company, to translate the fruits of university research into products and services that benefit the public.

The services offered by university TTOs vary somewhat, depending on the academic institution's size, research budget, culture, and local ecosystem. The common thread throughout is managing the institution's IP rights, mostly in the form of patents and copyrights; marketing and licensing those rights to industry to develop and commercialize useful products; and managing the distribution of income resulting from those licenses. More recently, many TTOs, particularly those of mid- to large size universities, have expanded their offerings to include one of more of the following: industry relations and industry sponsored research contracting, alliance management functions, entrepreneurship training, funds for translational research and proof of concept testing, startup services and seed funds, and accelerators and incubators. Many non-US institutions also offer to manage their researchers' outside consulting activities. The services offered by the TTO are especially important for therapeutic discoveries, considering 55% of FDA-approved NMEs originate from academia (Kim et al. 2017).

5.2.2 The Licensing Process

The invention management and licensing process generally begins with disclosure to the TTO of a detailed description of the invention and any relevant facts and circumstances, such as the list of inventors and any research sponsors. It is important for the inventors and the TTO to engage each other early in the process to make sure that everyone understands the various options for further development and commercialization, and the inventors and the TTO are aligned on the best path forward. If the invention does not need further maturation within the academic environment, the process will likely be fairly linear. More recently, as research teams move their biomedical inventions through various funding sources, translational programs, and/or industry collaborations, the process may have several touchpoints and can become more complex as some funders may impose IP or other obligations. The TTO will assess the commercial potential of the invention and may conduct an IP landscape analysis, after which a decision can be made whether to file for patent protection. While some software and digital-health inventions, as well as most biological research tools, will not require patenting in order to be commercialized, most inventions focused on therapeutics require patent protection in order to incentivize a company to develop and bring them to market.

Many TTOs have a technology marketing team that can help identify potential licensees and startup investors, although licensees are often identified through the inventors' own networks. If the technology requires further development and de-risking to attract external investment, an inventor-founded startup that initially makes use of a university accelerator or incubator and non-dilutive funding, such as government grants, may represent the best and possibly only avenue to bring a technology to market.

Once one or more appropriate licensees have been identified, the TTO will negotiate the financial and legal terms and conditions with the candidate licensee(s) or the startup entrepreneurs or investors. For startups, time-limited option agreements are generally less expensive and time-consuming than negotiating a full license agreement to secure the rights sufficient to get investors and management teams on board, after which a full license agreement makes sense. To mitigate conflict of interest concerns (see Sect. 5.3 for more on conflicts of interest), most universities will not negotiate license terms with a faculty member or student. Therefore, another entrepreneur or a lawyer should represent the startup company in the negotiations.

The research sponsors and all of the university stakeholders are strongly aligned in their desire to see new therapeutic technologies developed as rapidly as possible and in a manner that will lead to the greatest patient benefit. One of the functions of the TTO is to enforce licensee compliance with diligent development requirements, financial obligations, and other terms of the license. The TTO also manages the accounting and distribution of licensing income to various university stakeholders. Most academic institutions have policies that allow for a specified share of licensing income to be distributed back to the institution to support further research, as well as a personal share to the academic inventors.

> **Box 5.2: What Surprised an Academic?**
> Because the TTO is sensitive to the academic mission to share knowledge, filing for a patent rarely slows down publication, submission of an abstract, or giving a talk. Filing a disclosure with the TTO can be as simple as attaching your planned talk or draft manuscript to get the process going.—*DM-R*

5.2.3   Research Sponsor Rights

Biomedical research and development is an expensive proposition. Even within the university setting, millions of dollars in research funding may be needed in order to mature a therapeutic technology to the point at which it can attract external investment for the purpose of commercialization. With limited government funding available for the translational stage of research, universities rely more heavily on funding grants from non-profit foundations and charities, as well as from industry. These funding sources frequently seek some form of rights with respect to the resulting inventions. Industry sponsors will generally receive some sort of rights to IP developed under the funded project, often in the form of a time-limited option. It is helpful to note, however, that even non-profit funding grants sometimes come with legal obligations that cede some level of control over the IP rights and/or grant a specified financial return back to the funder. It is of utmost importance to fully understand and carefully consider the terms and conditions attached to both for- and non-profit research funding sources before accepting the funding so that any obligations under those grants are aligned with the development and commercialization goals for the technology. For example, some terms and conditions may adversely affect an academic inventor's ability to obtain future funding from other sources to further develop the technology or to form a new company to commercialize the IP.

5.2.4 Patent Management

Inventors should disclose inventions to the TTO as soon as there is data to support the conception, and in any case, several weeks before any public disclosures, including any publications in journals, conference abstracts, laboratory websites, and oral presentations in non-confidential settings. This will give both the TTO and patent attorneys time to work collaboratively with the inventors on when to file and what subject matter to include in order to best meet business needs and to prepare the application accordingly (see Sect. 5.1 for more on intellectual property). Many academic inventors are not aware of the high level of strategic thinking that goes into a patent filing, particularly for patents related to drug discovery. Timing and what information to include in the patent application are important considerations when determining the right balance between need to publish, meeting the first-to-file requirement, and ensuring that there is enough data in support of the desired claims to meet patentability requirements. Once the patent application is drafted, the inventors must work closely with the patent attorney to ensure that the full extent of the invention and its applications in terms of potential commercial products are captured in the patent.

The patent prosecution process leading to issuance of a US patent in the biomedical space generally takes three to five years, but sometimes longer. Patent filings can also become very costly, particularly if broad foreign rights are pursued, which can cost hundreds of thousands of dollars. Most TTOs have limited patent budgets and must carefully consider at various timepoints during the filing and prosecution processes which inventions warrant investment of funds for patentprotection and in which countries.

5.2.5 University Gap Funds, Accelerators and Startup Services

The ongoing challenges of obtaining seed funding for startups have increased the importance of further developing and de-risking new therapeutic technologies before exiting them out of the academic setting. Many academic institutions now offer gap funds, accelerator programs, and/or incubators to move novel, cutting edge technologies across the applied science stage that is often referred to as "the valley of death".

Gap funds and accelerator programs generally offer translational research funding, often in milestone-driven tranches, in combination with access to mentors and advisors who have expertise in drug development, and are familiar with current regulatory issues and other market forces (like SPARK!). Many of these programs also offer shared access to highly valuable drug discovery tools and equipment not generally available within academic laboratories. Incubators can also play an important role by providing cost-efficient access to quality wet lab space and shared business resources. Many incubators also have advisors on staff with expertise in drug development. Academics should familiarize themselves with and take advantage of the programs and services offered within their local ecosystem.

5.2.6   Conflict of Interest Considerations

Whenever university faculty found and take an ownership stake in a company to license and commercialize the university-owned IP that they invented, conflict of interest concerns will naturally arise. For example, is the faculty founder putting students on projects that will benefit the faculty member's startup rather than projects that will advance the students' academic goals? Is follow-on research being driven by the potential for personal gain (see Sect. 5.3 for more on conflicts of interest)? Faculty founders will often wish to negotiate the IP license on their own, but from a legal perspective, it may be difficult to determine whether they negotiated on behalf of the university as a university employee, or solely on behalf of the independent company.

With these and other questions in mind, most institutions with medical schools or academic medical centers have established mitigation mechanisms to maintain academic freedom, ensure that all research is conducted with utmost integrity, and protect students' academic goals and careers. Mitigation plans can include actions such as minimizing the personal equity holdings of the startup founders, providing full transparency to all affected lab members as to the existing financial conflicts, and creating oversight committees that track student progress and publications, among other things.

The Bottom Line
Technology transfer is essential for commercialization of discoveries and inventions from academia. The process can be complex and requires balancing the interests of numerous constituents, including inventors, licensees, the university, the government, and the public. The end result can be satisfying to all parties when university technology is successfully translated to industry for society's benefit.

5.3 Navigating Conflicts of Interest

Emily Egeler

As university faculty members get more involved in the drug development process, it is essential that academic institutions maintain their high-quality unbiased research and education missions. Faculty owe their primary allegiance to their institution and its core values. Investigators are also responsible to government agencies and the public who fund the research and ultimately to the patients who are treated by these new therapeutics.

The potential for financial conflicts of interest is common and sometimes unavoidable as academic investigators work to translate their discoveries. An increase in collaborations across academia and industry has arisen due to the push to move technology forward faster. The goal should not be to shun all interactions with for-profit entities, but rather to manage and report these interactions in a transparent way to protect the public.

Box 5.3: What Surprised an Academic?

It is not against most university and funding agency policies for academics to financially benefit from their research and clinical work. On the contrary, institutions should encourage faculty to work long and hard to make new therapeutics a reality. It is, however, essential that the public trust the impartiality of the research and clinical trial data generated in publicly funded institutions and that the independence of trainee education and mentorship be maintained. This is why it is so important to properly manage conflicts of interest.—*DM-R*

5.3.1 Identifying Conflicts of Interest

The Sunshine Act, which elicits transparency of physician payments or ownership interests made by all pharmaceutical, medical device, biotechnology, and medical supply manufacturers doing business in the US, went into effect March 31, 2013, as part of the Patient Protection and Affordable Care Act. Each institution has local policies for reporting financial interests and determining if the interests could have a direct and significant effect on the research. Currently, the Public Health Service, which encompasses many federal agencies including the National Institutes of Health (NIH), FDA, and Centers for Disease Control and Prevention (CDC), defines a significant financial interest as (a) income greater than $5000 per year, (b) equity or ownership in a non-publicly traded company, or (c) personal income from IP outside of any university paid royalties.

A conflict of interest (COI) depends on the situation and does not necessarily reflect on the character or actions of the individual. If a reasonable person might conclude that relationships or interests could influence academic responsibilities, investigators should disclose the financial interests following institutional protocols.

Box 5.4: Guide to Identifying Conflicts of Interest
1. Do any financial interests or relationships, including items without current value like stock options, relate directly to my responsibilities and role as an academic investigator?
2. Could my research findings impact the success of these interests, or appear to affect the success?
3. Could a reasonable person conclude that my research or role at the university might be influenced by my financial ties?
4. Does my institution have requirements to report interests for spouses, partners or dependent children?

Box 5.5: What Surprised an Academic?
Federal regulations and institutional policies may require us to report financial interests for our spouse, partner or dependents in addition to our own industry connections. Usually this is only necessary if their connections relate to our area of research or clinical practice.—*DM-R*

Disclosing a potential COI does not imply that the research or clinical work is actually biased; maintaining integrity of research findings is still a personal responsibility. Instead, disclosure of conflicts of interest is an avenue to increase transparency between financial interests related to the research and clinical outcomes in drug development. Managing a COI may be as simple as adding language to an informed consent form, hiring an outside statistician, or blinding the data to the investigator. Some financial interests are unmanageable COIs and will prevent participation in running or analyzing a clinical trial.

5.3.2 Conflicts of Interest to Avoid

Some industry relationships are almost never appropriate. Most importantly, clinical investigators should not hold primary responsibility for a clinical trial in which they stand to financially benefit from the outcome. Clinicians should attempt to minimize the possibility that industry relationships influence discussion of treatment options for patients who consider participation in the clinical study. For instance, clinicians should not accept personal payment for each patient they enroll or refer to a clinical trial.

On the laboratory research side, it is important that academics keep their university role and responsibilities separate from any for-profit ventures. For instance, graduate students or postdoctoral scholars should not be coerced into performing research that will benefit a faculty startup company or otherwise have their research topic areas restricted. Education must include openness in research with free exchange of scholarly information and opportunities to present work at seminars and in papers or dissertations without delay. An academic researcher should not accept corporate funding or in-kind gifts that require a delay in publication or that use university resources to preferentially benefit a particular company.

> **Box 5.6: What Surprised an Academic?**
> Faculty who serve on a scientific advisory board (SAB) or medical advisory board (MAB) may be compensated with stock or stock options (equity) in lieu of payment for services. Avoid accepting equity as payment if you are an investigator in related human subject research studies. Holding any equity can be construed as having the ability to compromise the objectivity of the research study. This includes stock options whose value cannot readily be determined.—*DM-R*

5.3.3   Managing Conflicts of Interest: Clinical Investigators

Disclosing conflicts of interest in the clinical research setting is intended to solidify public confidence in the results of clinical trials and to ensure that clinicians are acting in the best interest of their patients. Areas of concern include intentional or, more likely, unconscious bias in interpretation of study results, reporting of adverse events, or selection of a course of treatment. Accepting gifts from companies, including things like free drug samples or even pens with the company logo, is not allowed in some universities as such practices may potentially influence the open, independent environment of the academic institution or affect the university's tax-exempt status.

Clinical investigators should be able to partner with industry to improve patient care, but it is important to be aware of university policies and to avoid situations that might impact the impartiality of the study design, data collection, or findings. Expect to disclose conflicts of interest in relation to human subject research protocol submissions to the Institutional Review Board (or Ethics Committee), grant applications, or Material and Human Tissue Transfer Agreements. An institutional COI may arise when a university or hospital stands to benefit monetarily from clinical research results because of a financial relationship with a company. The potential for institutional COI should be considered when an institution is deciding whether to participate as a clinical site for a trial. A hospital or academic medical center may decline to participate if the institution would benefit financially (e.g., through future royalty streams, milestone payments, or equity ownership) from positive study results.

5.3.4   Managing Conflicts of Interest: Laboratory Researchers

The public relies on researchers from universities and other academic institutions to be impartial and comprehensive when reporting their findings. Not only is the public the end recipient of any drugs developed based on the research, but also the funders of research supported by federal grants. The most common financial

The Bottom Line

Conflicts of interest include any industry connections that could appear to influence the design, conduct, or interpretation of research or clinical trials. Disclosing a COI does not imply that the results are biased. Accurate disclosure of a COI is essential to maintain transparency and public trust in research. Consult university policies for specific guidance on what and how to report.

conflicts of interest for researchers arise from paid consulting, accepting monetary or in-kind support from an industry partner, or having equity ownership in a startup company that holds IP licenses from the university.

It is important that financial ties do not affect or appear to affect the design, conduct, interpretation, or reporting of research. Nor should industry affiliations interfere with the selection of research topics. Importantly, we must also protect the integrity of our educational mission. Our teaching and mentorship must not be biased by financial considerations. Agreements cannot restrict students' ability to publish or present results, and trainee work should be independent from any mentoring faculty's COI. Within the academic setting, all meetings should be educational and not for marketing purposes.

5.4   Working with the University Compliance Office

Jennifer Swanton Brown and Nicholas Gaich

This section focuses on academic institutional requirements for conducting human research, using Stanford University as an example. While each institution's compliance infrastructure and policies will vary to some degree, the guiding principles are the same: to protect human research subjects, to ensure high quality design and execution of research protocols, and to ensure that government regulatory requirements are met. We highly recommend that you consult your institution's policies before conducting human research.

Regulatory compliance at Stanford for research involving human subjects is defined in the Human Research Protection Program (HRPP) and administered by the Research Compliance Office (RCO) in the Office of the Vice Provost and Dean of Research. The RCO administers Stanford's Administrative Panels for Human Subjects in Research, the formal name for the Stanford Institutional Review Board (IRB). Panels for Human Subjects in Research committees in other countries have different names, such as Independent Ethics Committee (IEC) in Europe; Human Research Ethic Committee (HREC) in Australia, etc. Additionally, and based on allocation of funding for an investigator-initiated project, there may be university sponsored project and research compliance requirements administered by the Research Management Group (RMG).

For industry-funded clinical trials, a Stanford investigator may initiate all three phases of research oversight in parallel: IRB (human subject) approval, contract negotiation, and budget development. IRB approval is required for all human subject research and is not contingent on the other two processes. Detailed budgets are required for sponsored projects and strongly encouraged for department-funded projects. Contracts are contingent on meeting all sponsored project compliance requirements and IRB approval and are needed if drug, device, biologic, or funding is received from, or data, publication rights, or IP is shared with an entity outside Stanford, for example, from an industry collaborator or as a subaward from another academic institution.

5.4.1   IND Requirements

An Investigational New Drug application (IND) must be filed with FDA before any unapproved drug can be used in research with human subjects in the US. The Stanford IRB requires written evidence of FDA concurrence with the IND if the compound being studied has never received FDA approval for any use in the US.

If the drug being studied has prior approval by FDA for use in the US, the study may require a new IND or be IND-exempt. For instance, Stanford's IRB application walks the investigator through the regulatory questions that assist an investigator in making this determination. Clinical trials are typically IND exempt if the researcher studies the drug according to its legally marketed labeling. Trials typically require a new IND if the research involves a "route of administration or dosage level or use in a patient population or other factor *that significantly increases the risks* (or decreases the acceptability of the risks) associated with the use of the drug product" (Code of Federal Regulations 21 CFR 312.2(b)(iii)).

> **Box 5.7: What Surprised an Academic?**
> Clinical research with foods or nutritional supplements is an area that requires a nuanced understanding of regulatory affairs. An academic might reasonably expect that studying the effects of vitamins or herbal supplements to treat disease or alleviate symptoms is nothing like studying a new cell therapy or small molecule. FDA, however, considers foods and supplements to be drugs, if their research meets the regulatory definition of a clinical investigation. Navigating this regulatory landscape is complicated by the fact that such products are often commercially available, and their manufacturers may not fully understand the IND regulations. One investigator, following a nutraceutical manufacturer's advice, would have begun a clinical trial without an FDA-required IND but for the quick action of the IRB and expert knowledge in the Clinical Research Quality (CRQ) regulatory affairs team.—*JSB*

5.4.2   IRB Oversight

At Stanford, applications to the IRB are submitted via an electronic submission process, called eProtocol. For studies that are not IND-exempt, the IRB requires documentation of an FDA-approved IND. Acceptable documentation includes a letter issued by FDA indicating the IND number.

Informed Consent templates are available from Stanford's IRB website. The investigator needs to provide any advertisements or recruitment materials to the IRB at the time of application. Guidance on Data and Safety Monitoring or special regulations for research with children are also available among the IRB's comprehensive guidance documents. By carefully following the Informed Consent templates and fully answering the eProtocol questions prior to IRB application submission, a researcher can expect an approval from the IRB within 4–6 weeks. The Stanford IRB also reviews and approves Health Insurance Portability and Accountability Act of 1996 (HIPAA) authorizations for research.

There are additional requirements and approvals for research conducted at individual units within Stanford, such as the Veterans Administration Palo Alto Health Care System. Additional requirements and approvals may also be needed for certain indications, such as cancer, or for particular therapeutic modalities, such as stem cell research.

In the era of precision health and real-world data, additional reviews on data use, privacy or security domains, when needed, are integrated into the IRB system.

5.4.3   Sponsor–Investigator Research Training and Support

Sponsor–investigator research (SIR) is defined as research conducted by a Stanford investigator who holds an IND from FDA. The Stanford IRB requires protocol-specific sponsor–investigator research training for the Principal Investigator (and/or Protocol Director) and research team prior to initial IRB approval. This training is provided by regulatory staff from the Cancer Clinical Trials Office (CCTO) or Spectrum's Workforce Development program. Training includes review of required regulatory documentation and reporting to FDA and IRB. Prior to continuing approval of an ongoing project, RCO requires investigator self-assessments and conducts consent reviews. Individual departments and centers also provide oversight. Sample training and review documents can be found on Stanford's IRB website.

Training and regulatory consultation are available for any aspect of investigator-initiated research at Stanford (see Sect. 5.4 Resources for relevant websites).

- Sample templates, logs, and standard operating procedures (SOPs) are available through CCTO, Spectrum, and CRQ.
- Good Clinical Practice training is available upon request or when required.

- CRQ and CCTO regulatory staff assist with IND applications and amendments for submission to FDA, study initiation, processing and tracking safety reports, regulatory documentation, and IRB, SRC, and CTRU submissions.
- In the event of a regulatory audit, staff from CRQ and CCTO coordinate with auditors and the IRB, review study documentation prior to the audit, liaise with institutional officials, and provide expert advice.

5.4.4   Risk Assessment Committee (RAC) Review

Stanford policy requires industry partners to provide insurance for potential research-related injury during clinical trials they benefit from. However, industry increasingly chooses not to provide insurance coverage when a Stanford researcher conducts investigator-initiated research with an approved drug. In such cases, a waiver of this policy may be obtained by appeal to the Risk Assessment Committee (RAC). RAC approval is required prior to completion of the contract. Risk Management and the IRB also consult with RAC for studies with complex conflict of interest scenarios or involving high-risk first-in-human therapies.

5.4.5   ClinicalTrials.gov Registration

With the exception of phase 1 clinical trials, any study conducted with an IND requires registration with the ClinicalTrials.gov public registry. Registration must be completed prior to first patient enrollment. The IRB application triggers this process and guidance documents are available on the CRQ website. Generally, the principal investigator is responsible for registering their study and for meeting all the requirements of ClinicalTrials.gov registration, including uploading of study results.

The Bottom Line
Become familiar with your institution's policies, procedures, and support structures for the conduct of human research. They exist to protect you, your participants and patients, and your institution.

Recommendation(s)
Investigators who are contemplating IND research for the first time can obtain regulatory consultations from CRQ's regulatory affairs service. FDA "speaks a different language" that is unfamiliar to many academics, and investigators are well advised to approach the agency with an expert on the team. As investigator-initiated IND research has increased in complexity over the past decade; special circumstances requiring regulatory knowledge include the coordination of multisite studies, international sites, manufacturing quality management, and vendor relations.

5.5   Not-for-Profit Drug Development

Eugenio L. de Hostos, Grace Hancock, and Mary Wang

Statements about "not-for-profit drug development" (NPDD) are often followed by questions like, "Is there such a thing?" and "Does that even work?" The short answer to both questions is yes. NPDD for neglected tropical diseases (NTDs) and rare and other underfunded diseases is becoming more common due to a convergence of scientific, business, and social trends. Globalization has led to a raised awareness of the medical needs of the poor in the developing world and at the same time introduced new philanthropic, industrial, and governmental resources for those needs. However, the resources available for NPDD still pale in comparison to the concentrated and vertically integrated resources available to a conventional pharmaceutical company working on drugs for profitable indications. For this reason, the key to NPDD is the ability to harness scattered financial, material, and technical support from a variety of sources for a given project (Nwaka et al. 2009).

In the past two decades, not-for-profit organizations have emerged under different models to meet the health needs of the global population while working within the confines of pharmaceutical development. One model called Product Development Partnerships (PDPs) has emerged. PDPs such as Drugs for Neglected Diseases Initiative (DNDi), Medicines for Malaria Ventures (MMV), The Medicines Development for Global Health (MDGH), and the pioneering Institute for OneWorld Health (OWH—later the PATH Drug Development Program) have brought the rigor and expeditiousness characteristic of large pharmaceutical companies to NTD vaccine and drug development. The impact of PDPs is impressive, and one study suggested that some 75% of NTD drug development is now conducted by organizations of this kind (Moran 2005).

With the advent of high-throughput screening facilities in academic institutions and an increased interest in NTDs among students and professors alike, academia has become a breeding ground for early-stage drug development projects. Not only are more targets being validated, assays developed, and compounds discovered, but academics are increasingly involved in advancing these discoveries into drug products. However, many academic projects stall due to lack of expertise and resources when it comes to pharmaceutical development. PDPs can play a critical role in turning academic projects into drug development programs and shepherding development candidates through the preclinical "valley of death" (Emmert-Buck 2011; Moos and Kodukula n.d.).

As discussed earlier, screening hits and motivation alone are insufficient to bring a new drug to market. PDPs can be clever about picking and executing projects and can maintain lean operations, but there is no fundamental reason to expect NPDD to be cheaper than conventional drug development. Just like any biotech or pharmaceutical company, PDPs can keep their costs down through outsourcing drug development services. Organizations like OWH and DNDi have gone as far as becoming completely lab-less and thus "virtual"—focusing on project building and management while relying exclusively on the services of contract research organizations around the world for lab work and manufacturing.

As opposed to the lab-less PDP model, internal drug discovery divisions have formed within some larger academic and research institutes, with the capacity to develop products in-house. This lean business model combines tools, infrastructure, and expertise to carry out drug discovery ranging from large-scale screens to early-phase clinical trials in one institute. While initial efforts are carried out in a non-profit setting, the final products may be developed in the non- or for-profit sector. For instance, Calibr, the drug discovery division of Scripps Research, furthers the development of drugs spanning both NTD and more profitable indications with high unmet need. Monetary, structural, and technological support for the for-profit projects may come from partnerships with large biopharma companies, CROs, computational platforms, and joint ventures launched in partnership with venture capital (VC) funds. These funding streams make it possible to hire and retain drug development experts, who also contribute to the non-profit projects. Nonetheless, finding development partners when non-profit programs are ready to leave the university or research institute can be challenging.

Repurposing drugs is another cost-saving option for PDPs. OWH, for example, repurposed paromomycin, an off-patent antibiotic approved for human use decades earlier, to treat visceral leishmaniasis by conducting the requisite phase 3 studies. Similarly, the veterinary pharmacopeia, particularly in the field of antihelminthics, is a good place to look for drugs with "crossover" potential; these can be developed for human use more quickly and at a lower cost than starting from scratch (Olliaro et al. 2011). Calibr and the Bill & Melinda Gates Foundation built ReFRAME, a best-in-class collection of 14,000+ known drugs. Through this chemical library, high-value assays can be deployed, especially in challenging therapeutic areas that have a paucity of leads or difficult assays. To date, Calibr with the Gates Foundation has discovered numerous opportunities for both direct repurposing (e.g., auranofin, clofazimine) and accelerated lead optimization programs to generate fit-for-purpose new chemical entities (NCEs). It is important to keep in mind, however, that repurposing still requires phase 3 studies at the very least, which are usually the most expensive part of drug development. Another important consideration regarding repurposed drugs is that in the end, the new product may fall far short of the ideal target product profile. For this reason, the financial advantage of repurposing has to be weighed against the long-term utility of the product.

On the income side, the situation is very different and more challenging for NPDD than for conventional drug development. As long as not-for-profit drug development remains not-for-profit, philanthropy will remain the "secret sauce" that makes NPDD possible. It is basic capitalism: social goods that will not generate sufficient profit to motivate the commercial sector require funding by government and philanthropic agencies instead.

Grants from organizations such as the Bill & Melinda Gates Foundation (see Sect. 5.5 Resources), the Wellcome Trust, and the World Health Organization's Special Programme for Research and Training in Tropical Diseases (TDR) are the lifeblood of many PDPs, but in-kind donations from the pharmaceutical industry are also very important. For example, a number of pharmaceutical companies have provided generous support to OWH's Diarrheal Diseases Program, including technical advice, access to compound libraries, drug discovery, and preclinical services. That said, the mantra of PDPs is to become less dependent on philanthropy and find a way to become more sustainable.

Financial sustainability not only allows long-term strategic planning, but also provides greater flexibility in incubating early-stage projects and exploring new development possibilities. To achieve this, options include partnering externally, building internal resources, or something in between. One strategy commonly used to help sustain PDPs is to include in its development portfolio a drug product with dual markets—that is, a profitable, rich-world market and a not-profitable, developing world market. For example, a drug to treat cholera that could also be sold to treat traveler's diarrhea could help support other projects in a PDP's pipeline. In the area of vaccines, Advanced Market Commitments (AMCs) are being used to incentivize the development and production of new products. An AMC is simply a promise by an organization (e.g., World Health Organization or government) to purchase a certain amount of vaccine product at a certain price once development is complete. This mechanism has been used to incentivize vaccine manufacturers to make flu vaccines and is now being used to promote NTD vaccine development.

Another small but wealthy market that can potentially support the development of NTD drugs is the military. For example, the US Department of Defense supports drug development in areas of military and national security importance. In some cases, these include NTDs (see Sect. 5.5 Resources). PDPs can also benefit from a range of government resources and incentives such as those offered by NIH and FDA. These resources are spread out and each alone is unlikely to support a single development phase in its entirety, let alone a complete drug development program. However, an NPDD program can be greatly aided by access to these available resources, though timelines will likely be extended. The National Institute for Allergy and Infectious Diseases (NIAID)'s Division of Microbiology and Infectious Diseases also offers drug development services ranging from discovery to phase 1 trials through government contractors (see Sect. 5.5 Resources). NIH also offers a program called Therapeutics for Rare and Neglected Diseases (TRND) that makes available the expertise and resources of institutes such as NIAID, in effect establishing an ad hoc PDP with the applicant (see Sect. 5.5 Resources). TRND, for example, serves as a portal for resources at NIAID earmarked for work on schistosomiasis and hookworm disease. Services offered through the TRND umbrella are also available to applicants outside the US.

On the regulatory side, FDA has developed several programs intended to encourage development of treatments for diseases neglected by the pharmaceutical industry (see Sect. 5.5 Resources). The Orphan Drug Act was enacted to reduce the costs of developing drugs for diseases that affect less than 200,000 patients per year in the US or drugs not expected to recoup their unaided development costs, both categories that include NTDs. The Office of Orphan Products Development (OOPD) is responsible for facilitating orphan drug development by finding regulatory shortcuts that can save time and money. One of the mechanisms that the OOPD promotes is repurposing of existing drugs from the human and veterinary pharmacopeia.

Another program that has generated great expectations is FDA's Priority Review Voucher (PRV) program, mandated by US Congress in 2007 to promote drug development for less profitable indications. Under the program, any organization that obtains approval of a New Drug Application (NDA) for an NME to treat a disease on its list of NTDs is granted a transferable PRV (see Sect. 5.5 Resources). In 2012, this program was expanded to include rare pediatric diseases, and in 2016, for material threat medical countermeasures. The potential value of the voucher lies in the fact that it gives the bearer the right to an expedited review of another NDA application for a drug that need not be to treat the disease for which the voucher was awarded. Obviously, this could incentivize a big pharmaceutical company to take a second look at shelved projects that may have potential to treat one of these government priority areas. However, what excites the PDP community the most is that PRVs can be sold to a third party, such as a pharmaceutical company seeking to shorten the time-to-market of its new blockbuster. The proceeds can then be used to support additional non-profit drug development programs.

To date, 10 PRVs have been awarded for therapeutics for neglected diseases (three others have been awarded for vaccines, and separate programs exist for pediatric rare diseases and for medical countermeasures), and two of these have been sold for close to or more than $100 M (see Sect. 5.5 Resources). It is important to note, however, that all of these molecules were discovered before the PRV program was launched and, in fact, some were already in use in veterinary (moxidectin and triclabendazole) or clinical practice (miltefosine, nifurtimox, and benznidazole) years before the PRV program, though they were not FDA-approved. That said, the PRV program is an important incentive that already has had a positive impact in the development of drugs for neglected diseases and is likely to continue to have one in the years ahead.

5.5.1 Conclusion

Bringing new drugs to market is a complex and expensive process, but not-for-profit drug development is possible, is happening, and is arguably more feasible than ever before. In this resource-constrained corner of the pharmaceutical world, Product Development Partnerships and other non-profit organizations will continue to take a leading role in effectively harnessing a wide range of scattered public, industrial, academic, and philanthropic resources to fill in gaps in the treatment of neglected diseases.

The Bottom Line
Academics can advance their drug development projects for diseases of the developing world by partnering with a philanthropically funded Product Development Partnership or government program, such as NIH's TRND.

Resources
- Bill and Melinda Gates Foundation. Global Grand Challenges. Grant Opportunities: https://gcgh.grandchallenges.org/grant-opportunities
- Bill and Melinda Gates Foundation. Requests for Proposal Opportunities: https://submit.gatesfoundation.org/
- Department of Defense. US Army Medical Research and Development Command: https://mrdc.health.mil/
- Department of Defense. Defense Advanced Research Projects Agency (DARPA): https://www.darpa.mil/our-research
- NIH National Institute of Allergy and Infectious Disease (NIAD). Division of Microbiology and Infectious Diseases: https://www.niaid.nih.gov/about/dmid
- NIH Fogarty International Center: https://www.fic.nih.gov/
- NIH National Center for Advancing Translational Sciences. Therapeutics for Rare and Neglected Diseases (TRND): https://ncats.nih.gov/trnd
- FDA Tropical Disease Priority Review Voucher Program: https://www.fda.gov/about-fda/center-drug-evaluation-and-research-cder/tropical-disease-priority-review-voucher-program
- FDA Orphan drug designation: https://www.fda.gov/industry/medical-products-rare-diseases-and-conditions/designating-orphan-product-drugs-and-biological-products
- Duke University. The Fuqua School of Business. Priority Review Vouchers: https://sites.fuqua.duke.edu/priorityreviewvoucher/

5.6 Personal Lessons from a Reluctant Entrepreneur

Daria Mochly-Rosen

In 2001, my graduate student Leon Chen was concerned that our inhibitor of delta protein kinase C (δPKC), which dramatically reduced cardiac damage after a myocardial infarction, did not receive any interest from the pharmaceutical industry. It was inexplicable to us. We had found and published that delivering this short peptide (δV1–1) immediately after heart attack reduced 70% of the damage to the heart in mice, rats, and pigs and prevented heart failure, a common outcome in humans who suffered heart attack. Independent researchers even showed similar results in rabbits and guinea pigs. Furthermore, we had published a great deal regarding the molecular and cellular actions of δV1–1 and δPKC. And the market—the unmet patient need to prevent damage from heart attack—was and remains really large. Stanford's Office of Technology Licensing had patented δV1–1 and the information related to the discovery. How could the pharmaceutical industry not recognize its value?!?

My effort to introduce δV1–1 to the pharmaceutical industry was a total failure. Although I had a chance to speak with top decision makers in some of the large pharmaceutical companies, I was firmly turned down. Some of them provided specific feedback: "it is unlikely to work because we never saw an agent like that, a target like that, a time to intervene that you used" and so on. Although we provided very substantiated data (even vis-a-vis Glenn Begley's rules; see Sect. 3.2), they felt that it was too risky.

When Leon graduated, we began a campaign to raise money for our startup, KAI Pharmaceuticals. I was very reluctant to do so because my heart was in basic research, but I believed in this drug. The following is a short list representing some of the many lessons that I learned.

- Lesson 1: Consult, Consult, and Consult More!
  There was nothing in my academic background that prepared me with the know-how required to found a startup. I bought a notebook and wrote a list of questions for which I sought answers from various experts. Luckily, the first person I spoke with was the late Professor Ken Melmon of Stanford. He suggested that at the end of each meeting I should ask "is there anything else I should have asked?" something that I continue to do to this day.
- Lesson 2: Repeat What You Hear
  The challenge in talking with experts outside your field is not when they use a term that you have never heard before, it is when they use one common to your field, but with a different meaning. Even if you do not speak Italian, you understand the meaning of "stupido" or "intelligente"; you would know to ask what does "seggiola" mean, but would you know to ask what "caldo" means? If not, you will be in "hot" trouble (caldo means hot not cold). The trick is to repeat what you hear in your own words.

- Lesson 3: Leave Your Ego at the Door
  Pitching your life's work to sometimes uninterested and otherwise inattentive VCs who may have less knowledge or achievements than you can be frustrating, especially if they turn you down. However, the pitch is not about you, but about the patients that you will impact with the solution that you build. So why be hurt by the rejection? In every pitch and interaction, you likely will hear good questions and suggestions. You should use these to improve your next pitch—after all, all you need is one "yes."
- Lesson 4: Don't Put All Your Eggs in One Basket and Keep Some of Your Powder Dry!
  To reduce the risk for your company, it is really important to invest in plans B, C and even D, as backups. Things never go exactly as planned and some good ideas will hit a wall. It was important for KAI's success that we maintained an active research group, even though the company had entered expensive clinical studies. It paid off that we resisted putting all our resources behind what seemed to be the winning horse. (See next Lesson.) And back to lesson 2, if like me, you do not know what keeping your powder dry means, just ask.
- Lesson 5: When Failing, Think Before Giving Up!
  This lesson I learned from John Walker (see Sect. 5.12) who was our CEO when one of our programs failed a phase 1 clinical study because of a severe side effect (hypocalcemia). John organized a party to congratulate the team for the excellent work that led to a "no-go" decision. Although none of us were in the mood for a party, it taught us that we were a team through thick and thin. And it led us to discuss whether there was a clinical use for a drug that caused hypocalcemia. The basic research team took on the hard work to examine the molecular basis for the effect. The finding that it bound the calcium sensor in the parathyroid gland suggested that it could be used to treat secondary hyperparathyroidism associated with kidney failure. This ultimately resulted in KAI's acquisition by Amgen and a novel drug on the market—Parsabiv.
- Lesson 6: Know Your Market; It Keeps Changing
  Dr. Kevin Grimes, the medical director of KAI, and I wrote a commentary about why eventually δV1-1 did not hit the market (Mochly-Rosen and Grimes 2011). Heart attack continues to be a common disease (800,000 people have a heart attack each year), but the addition of multiple drugs (aspirin, thrombolytics, platelet inhibitors, beta-blockers, angiotensin-converting enzyme (ACE) inhibitors, etc.) to our therapeutic armamentarium has greatly improved outcomes. When we first planned our clinical studies in 2002, about one in six of the patients undergoing angioplasty for a heart attack died or developed heart failure. By 2006, these combined outcomes decreased to around one in 25. Heart attack is still a very large market, but in order to show impact on clinical outcomes, we needed to enroll many more thousands of people, which was prohibitively expensive.

- Lesson 7: The Team's Strength = the Weakest Link!
  It is easy to pay attention to the work of the scientists and key medical officers and forget that all the other employees are also critical to ensure the company's success. At KAI, we had many "all hands on deck" events, especially when it felt like we had no time to waste. Everyone knew one another well and all were included in meetings reporting on the company's progress as well as in games, outings, and barbecues. That probably explains why one early Monday morning I found the person in charge of washing the animal cages resting in the board room. He had worked through the previous night to wash the cages by hand, because the cage washer broke. No one called him in, but he decided to come in and check, because he knew that an important shipment of animals was due to arrive that morning. When all members of the team are informed and feel valued, it pays off.

I learned many additional lessons during my time at KAI, sometimes by making unfortunate mistakes and often by learning from my many mentors. Entrepreneurship is really hard and no one person knows all that needs to be done. Remain open, ask questions, and continue to learn. And when you can, share lessons learned with others.

5.7   Selecting the Market for Your Drug

Robert F. Booth

Developing a new drug is expensive. Marketing a new drug is very expensive. Clinical trial success rates are low. Furthermore, for any potential new drug, there are often several different clinical indications and market opportunities that could be selected. Given these challenges, it is very important to carefully consider each of these parameters to ensure that the correct patient population and optimal target market is selected for a potential new drug.

5.7.1   The Target Product Profile and Selecting the Market Opportunity

A target product profile (TPP; covered earlier in Sect. 1.4) provides the blueprint for the desired profile of a potential product that is aimed at a particular disease or diseases. In designing the TPP, it is imperative to understand the target patient population and the market opportunity for the potential product, and not simply what the optimal profile of the potential product may be at the end of the preclinical phase of development or even at the end of phase 1 or phase 2. An understanding of existing and future medical needs of the target patient population and existing gaps in medical treatment will help frame the market opportunity. The position of existing competing drugs and those that may emerge in the future will directly influence what will constitute the market opportunity. Therefore, a robust and clearly defined TPP is needed to aid rational clinical trial design decisions and to provide a stronger business case for the development and future commercialization of the potential product. Since drug development is such a capital-intensive business, the vast majority of potential investors will want to evaluate the TPP(s) that the new company has designed. This should be viewed as an opportunity to persuade potential investors of the merits of the newly designed product.

5.7.2　Determining the Existing Market Size

Establishing the potential market size for a drug is a critical parameter as it helps to determine how much value a new product might represent. Most potential partners and investors will want to understand what the market size for a potential new product might be, and this will be a consideration in how much value they will attribute to a new startup company. The market size is broadly defined by how many patients might use a new therapy and how much revenue will be obtained by selling the new therapeutic. A simple way to identify the number of potential patients is to first evaluate the prevalence and incidence of the disease state that will be treated. Prevalence refers to the number of people in a population who have a disease or other health outcome at a given point in time. Incidence refers to the number of people in a population who develop a disease or other health outcome over a period of time (i.e., new cases during that period of time). As a broad guideline, it is simplest to use prevalence numbers to evaluate new therapies that will be used by the same patients on a recurring basis. For example, this could be patients who take an antihypertensive to treat high blood pressure. Incidence numbers are often best used to apply to products that will be used to treat a single, acute event, such as a stroke or an acute bacterial infection.

Since not all potential patients will be treated with a new therapeutic, it is important to estimate the addressable market size. This number will be strongly influenced by the percentage of patients that will be diagnosed with the disease and the percentage of patients that will then be treated for the disease.

Box 5.8: Factors That Determine Market Size
- Prevalence and incidence of the medical condition
- Diagnosis Rate: How many new cases are diagnosed each year?
- Treatment Rate: Of those diagnosed, how many patients seek treatment?
- Patient Adherence: Assess likelihood of initiation of treatment, implementation of treatment, and discontinuation of treatment
- Patient Subpopulations: Are there subpopulations of patients who may show enhanced efficacy or greater adverse events in response to the treatment? Is there an opportunity for personalized medicine?
- How readily will the new treatment penetrate the existing market?
- Will the new treatment be best-in-class or first-in-class?

5.7.3 Patient Adherence

A further key modifier of market size is represented by patient adherence to the new therapy. A consensus published in 2012 characterized adherence to medications by three major components: the initiation, the implementation, and the discontinuation (Vrijens et al. 2012). Many potential factors can impact adherence, which can be separated into five broad categories: sociodemographic (e.g., income level, health literacy); healthcare system (e.g., patient-clinician communication, access to and cost of care); characteristics of the therapy (e.g., complexity of dosing, adverse effects); condition-related (e.g., symptom severity, impact on quality of life); and patient-related (e.g., perception of severity of illness and efficacy of treatment, lack of follow-up). In the literature, good adherence has been defined with a cutoff of 80%, although the relevance of this number has been questioned (Burnier and Egan 2019).

5.7.4 Patient Subpopulations and Personalized Medicine

For many disease indications, there will be subpopulations of patients whose response to a given treatment may yield enhanced efficacy or greater adverse effects. These subpopulations may be defined by a variety of criteria including age, sex, genetics and ancestry (see Sect. 3.5 for more on pharmacogenomics), disease severity, and disease characteristics. In instances where significantly different efficacy or adverse event responses are elicited by a given treatment, it is important that the appropriate subpopulation is targeted. Targeting the appropriate subpopulation during clinical trials should result in improved response rates, which could reduce the size and duration of the required trial needed to show efficacy and safety. If such a trial design provides evidence that the new product is markedly superior to competing products, premium pricing for the new product may be possible, which would increase the potential market size for the product. Note that targeting a subpopulation within a given disease indication will reduce the number of patients available to conduct a clinical trial and will reduce the overall number of patients that might have access to the new product. However, given the many advantages that may accrue from targeting the appropriate subpopulation, this parameter should be a major consideration regarding market potential.

Over the past decade, targeting patient subpopulations has been reflected in the use of personalized medicine that uses an individual's genetic or biomarker profile to guide decisions regarding the prevention, diagnosis, and treatment of disease. This is most clearly seen in the emergence of targeted cancer therapies that have been completely revolutionized, as several molecular alterations have been identified as drivers of cancer development and progression. An example of this approach is the use of pembrolizumab, sold under the brand name Keytruda. Keytruda is a monoclonal antibody that binds to the checkpoint inhibitor, programmed cell death 1 (PD-1), which is expressed on T cells. When Keytruda is bound, it prevents PD-1 from binding to ligands PD-L1 and PD-L2 expressed by certain cancer types and therefore prevents the tumor from downregulating the activity of T cells. As a single agent, Keytruda is indicated for first-line treatment of patients with non-small cell lung cancer expressing PD-L1, as determined by an FDA-approved test, with no epidermal growth factor receptor or anaplastic lymphoma kinase genomic tumor aberrations. The prevalence of PD-L1 expression in patients with non-small cell lung cancer ranges from 24% to 60%; therefore, setting a cutoff for positivity at 5% (Yu et al. 2016) could reduce the potential market for Keytruda in this indication.

Although cancer is an obvious target for personalized medicines, as new bio-marker technologies continue to emerge, this approach may be usefully applied to almost all therapeutic domains in which the drug target shows genetic polymorphism.

5.7.5   Market Penetration

Another modifier that helps determine market size is the ability of a new therapy to penetrate an existing market. Some guidance for this parameter may be obtained from prior penetration rates of comparable products. The comparative efficacy, ease of use, and safety of the new therapy compared with competitor products will strongly influence market penetration; both existing competition and potential future competition should be considered. The pricing of a new product will also influence its market penetration. One consideration that should be added to this analysis is the potential for generic products to enter the marketplace, since such generics will usually be priced very low compared with a brand name drug. Interviews with relevant physicians and payer groups can provide insight into the ability of a new product to penetrate a market. It is worth noting that market projections are frequently overestimated.

5.7.6   First-in-Class or Best-in-Class

It is also important to determine whether the new potential therapy will be a first-in-class or a best-in-class. These two parameters are clear determinants of future commercial success and should therefore be an essential consideration when constructing a TPP. First-in-class molecules are those that use a novel and unique mechanism of action to treat a medical condition, while best-in-class offers a clinical advantage over what is currently available on the market for a given indication. An example of a best-in-class molecule that was a late entry to the market was Lipitor, the HMG-CoA reductase inhibitor that was the fifth statin to reach the marketplace and was launched nine years after the first statin. Despite its late launch, Lipitor captured double the peak annual sales of other statins. This might suggest that it is more important to be best-in-class rather than first-in-class. However, an excellent analysis of drug launches in the 1990–2010 period by Shulze and Ringel (2013) indicated that it is slightly better to be first, rather than best (Schulze and Ringel 2013). The analysis showed that the value of late entries to the market falls dramatically, if the second entry is not rapid. The analysis highlighted additional factors that may significantly modify the first-in-class/best-in-class guidelines: (1) If the compound can target a patient subpopulation where it shows certain key advantages; (2) if the use of the drug can be expanded to additional therapeutic indications beyond its initial approval; (3) if the drug targets a large market, where different drugs might be cycled as part of the treatment regimen of a patient; and (4) if the drug is being marketed by a formidably strong commercial organization. Overall, however, the timing of market entry and actual therapeutic advantage remain the dominant parameters determining commercial success and these two considerations should remain pre-eminent in the TPP.

5.7.7   Selecting the Appropriate Therapeutic Area

For a given research project, there may be several different options for diseases to target. It is worthwhile to consider these options in terms of the costs that might be incurred to bring such a therapy to the market, since these vary significantly depending on the therapeutic area. A recent study indicated that between 2009 and 2018, the mean capitalized research and development investment to bring a new drug to market was estimated at $1335.9 million. These costs were highly dependent on the therapeutic area in which the drug was being developed. For example, median estimates by therapeutic area ranged from $765.9 million for nervous system agents to $2771.6 million for antineoplastic and immunomodulating agents. Furthermore, several studies have indicated that clinical trial success rates from phase 1 to FDA approval varied widely by therapeutic area, with the highest success rate in ophthalmology with a phase 1 to FDA approval rate of 32.6%, whereas in oncology, this rate dropped to 3.4% (Wouters et al. 2020).

Box 5.9: Some Questions for Target Groups
1. For patients:

- What is the current treatment for their medical condition?
- Are patients satisfied with current treatment options? And if not, why not?
- Is their quality of life improved by their existing treatment?
- Do they fully adhere with the treatment regimen as recommended by their physician?
- Is the dosing schedule or formulation inconvenient?
- What would they like from a new treatment?

2. For physicians, nurses and caregivers:

- What is the current standard of care that they use?
- What challenges exist in treating this disease?
- Are there dose-limiting interactions with other drugs?

- Where is the current treatment administered (at home, hospital)?
- Are patients readily able to access the treatment?
- Is the prescribing physician a specialist or a primary care physician?
- Is recruitment and/or patient retention difficult within this patient pool (e.g., competing trials, non-adherent subjects)?
- Are there subpopulations of patients who will respond optimally or adversely to treatment?
- How costly would a trial be in this indication (size, duration, endpoints)?

3. For hospital administration/payers:

- How costly would a trial be in this indication (size, duration, endpoints)?
- Are there challenges for reimbursement?
- What is necessary to prove cost-effectiveness in clinical development?

The Bottom Line

Clearly, the dominant parameter determining the selection of the optimal clinical indication should be the science that underpins the research driving the development, but careful consideration of the advantages and disadvantages of different therapeutic options should enhance the ability to fundraise and provide future partnering opportunities.

5.7.8 Parties to Be Considered When Identifying Unmet Medical Needs

The most obvious customer for a potential new therapeutic is the patient that will hopefully benefit from the new treatment. However, physicians, nurses, and other caregivers will also have significant input on how and whether the new treatment will be adopted. Additionally, those that might offer reimbursement for the new treatment, for example, a health insurance company, will also influence whether a treatment will be utilized. Similarly, hospital administrators will also have a significant input in deciding whether a new treatment will be used. Some of the questions that each of these groups will consider are listed in Box 5.9.

> **Box 5.10: How Many Potential Markets Should You Analyze for Your Drug?**
> If your drug may be used for multiple indications, evaluate only two to three markets and ignore the rest; evaluating each and every potential market in which a drug could be used is unnecessary. Most pharmaceutical companies believe in gated investing, which means that the drug must show promising activity in the first few indications before more investments are made. Focus on evaluating the first two to three indications where the product could launch and then list areas for future exploration without valuations. Notably the first indication does not have to be the largest market. Gated investing allows you to create a balance between market size and time/risk of demonstrating clinical utility. It will save you time and give you credibility.

5.8 Commercial Assessments

Julie Papanek Grant and Leon Chen

Companies, inventors, and investors complete commercial assessments to estimate the future revenue of a potential new therapy. Revenue from product sales provides funds for research and discovery investments, drives investor interest, and finances the delivery of products to patients. During the R&D stage of a product, commercial assessments inform whether the expected revenue is likely to provide an attractive financial return on the millions of dollars invested in the drug development process. Although it is unlikely that you will have to create a highly detailed commercial assessment as a translational scientist, identifying the key drivers will enable you to understand the market and product characteristics that make companies and investors take notice.

5.8.1 The Formula

Most pharmaceutical companies use the same formula when estimating annual sales:

$$\text{Market size}\left(\text{number of potential patients}\right) \times \text{Product share}\left(\%\text{taking your product}\right)$$
$$\times \text{Price per patient}\left(\text{paid for your product}\right) = \text{Annual sales revenues}$$

Think of the formula like a pie chart. The market size is the entire pie. The product share is the size of your product's slice expressed as a percentage. Price per share estimates the dollars that will be charged for the product and converts units into dollars in a given year.

This simple formula makes it easy to see trade-offs between different commercial and development strategies. Some companies prefer to pursue very large markets where there are lots of competitors because even a small slice of a big pie will provide a large return. Markets for statins, anticoagulants, and angiotensin receptor blockers are good examples. Other companies might choose to develop products for smaller markets with high unmet medical needs, few competitors, and few pricing constraints. In these markets, a larger product share and higher price per course of treatment can offset a small patient population to ensure adequate revenue.

Companies and investors seek to maximize the commercial potential of their products. Decision-makers focus on the variables that can be manipulated or influenced and have the largest impact on value. We will explore each of these drivers, so that we can see where drug development decisions impact commercial value.

5.8.2 Market Size

The market size is the number of patients with a specific disease who could be treated with a product in a given year. Start with the number of patients who are diagnosed with the disease each year. This initial number can then be adjusted up or down to reflect potential patient subpopulations or changes in diagnosis rates, treatment rates, the incidence of the disease, and the product's expected FDA label statement. When estimating changes in incidence, diagnosis, and treatment rates, consider external factors such as the (in)convenience, cost, and behavioral changes your product may impose upon physicians, nurses, and patients. These considerations are discussed more in the previous section, Sect. 5.7.

The expected FDA label statement is directly within the control of the team designing the clinical trials, in particular the pivotal trial(s), for a new product. The eligibility criteria for patients participating in the phase 3 studies will inform the FDA approved indication statement, which then drives which patients will receive the new product once it is approved. Diagnostic tests, drug–drug interactions, minimal vital sign requirements, age, ancestry, gender, and other patient characteristics can all dramatically change the number of patients who can receive a product. Each of these constraints should be considered when estimating the market size.

The inclusion criteria not only impact the commercial value but also the trial size, timelines, and probability of meeting the primary endpoint. Larger markets tend to require larger numbers of enrolled patients and more expensive clinical trials. As a result, trade-offs in phase 2 and phase 3 trial design should be debated by diverse teams including clinicians, biostatisticians, and commercial representatives.

> **Box 5.11: Real World Example**
> When estimating the market size for crizotinib, a drug approved for anaplastic lymphoma kinase (ALK) positive non-small cell lung cancer, Pfizer needed to account for the drug's specificity. Fewer than 2000 of the 221,000 patients diagnosed with lung cancer each year actually harbor the ALK mutation and are expected to respond to crizotinib. When FDA approved the drug in 2011, its use was restricted to patients with advanced stage non-small cell lung cancers that express the ALK gene. As a result, the FDA label reflected an approved market size of 2000 patients; not 221,000 patients.

5.8.3 Product Share

The product share estimate is usually the most subjective and heavily debated aspect of a valuation. Market share predicts the percentage of patients expected to be prescribed your product. In the end, most companies are looking for products that are first to market and/or the best drug on the market versus the competition. Better products and very novel products tend to have greater product share. Competition includes any and all therapies or procedures that can be used to treat a disease, not just drugs with the same mechanism of action. Surgical procedures and lack of treatment should also be considered as competition. Do not just include marketed products, because products in late-stage development (phase 3) and products that are similar to your product but ahead in development are all competitively relevant.

Speed and differentiation are critical components to determining product share. When a product is launching a year or more after a similar competitor, the second product will have a much harder time displacing the entrenched market leader. Being first to market carries the advantage of establishing a base of patients that are well served. As a result, the second product will often target patients not served by predecessors or seek to take market share by virtue of being better than the incumbent. Speed also influences the duration of patent coverage following market launch. Most investors want at least 5 years of patent protection after launch to allow time to recoup development costs before facing competition from generics or biosimilars.

Differentiation is a product's uniqueness in comparison to other options available to treat a specific disease or condition. It is relatively easy to imagine products differentiating on efficacy or safety, and indeed most second-in-class products aim to improve on one or both of these parameters. However, differentiation can be much more nuanced. Characteristics such as route of administration, dosing frequency, and contraindications can make meaningful differences in capturing market share. Other factors such as supply chain, storage conditions, and shelf-life can be competitive points of differentiation as well. While there are multiple ways to distinguish your product from predecessors, it will be important to compare all features of the different products in totality. Depending on the disease and treatment alternatives, patients and providers will probably show little tolerance for trading convenience for reduced safety or efficacy.

Finally, keep in mind payers' priorities when calculating product share. Consider whether your product is more cost-effective or improves patient outcomes over competitors. Make sure your reimbursement strategy at a high level is not a hurdle to physicians and patients adopting your product. As you move closer to launch, a detailed reimbursement strategy will be required.

By combining speed, differentiation, and competition, you will be able to make an educated guess about the product's market share. As a rule of thumb, a product will only be broadly adopted if it is better than existing options based upon criteria that matter. If a market has many undifferentiated competitors, speed matters most. Therefore, it is not surprising that companies pursue those therapies that could be first-to-market or best-in-class.

Box 5.12: Deciding on a Market Share Percentage

Weighing the importance of speed, differentiation, and number of competitors is highly subjective. There is no right answer. Similarly, precision down to the exact percentage point is not the goal. Instead, forecasters just want to ensure they are in the right range (0–15%, 15–30%, 30–60%, 60–80%, 80%+). The best evidence to support an assumption is analogies from similar products and similar markets. For instance, if a company is developing a rheumatoid arthritis therapy that will likely have lots of competitors, examples from the TNFα inhibitor market would be informative.

Box 5.13: Guidelines for Thinking About Market Share

1. Current options:

   - Are physicians, patients, and payers satisfied with the current options available?
   - What criteria would motivate them to switch to your product?

2. Similar products:

   - How many other products are being developed in the same market with the same or similar mechanisms of action as your products?
   - How is your product better?

3. Other novel approaches:

   - Approximately how many other products being developed in the same market have a different mechanism of action?
   - Which are most promising?
   - When will they launch?
   - How is your product better, the same, or worse?

5.8.4 Price

The final step in the formula is converting the number of patients treated with your product into sales revenues. A rough estimate of revenues can be calculated by multiplying price by number of patients. "Price" can encompass total cost to insurance companies per patient per year. This generalization accounts for different therapies with variable dosing schedules and frequency.

Whereas the dosing schedule and frequency will likely be fixed in the clinical design, the pricing estimate is unquestionably flexible. The same competitive assessment of speed, differentiation, and competition that informs product share also informs pricing. If a product meets a significant unmet medical need and dramatically improves outcomes or standard of care, a higher price (a.k.a. premium price) can be charged and will likely be reimbursed by insurance companies. Other improvements in safety, convenience, patient adherence, etc., typically do not command premium pricing.

Leveraging existing pricing estimates from the current standard of care is a good starting point. Similar to demonstrating clinical benefit on top of standard of care in a clinical trial, payers will compare price to standard of care options. If predecessors do not exist because the product is first-in-class, comparable pricing is more challenging. You may need to extrapolate from unrelated markets, but similar profiles. For example, many enzyme replacement therapies for different orphan diseases have similar pricing models despite treating distinct diseases. Gene therapy and CAR-T therapy are other analogous examples.

5.8.5 Making Informed Decisions Early

Commercial assessments should be incorporated at the earliest stages of a drug discovery project. Conviction in the fundamental biology and the ability to drug a target and impact disease is insufficient to achieve the goal of delivering a therapeutic solution to impact public health. A good commercial assessment will help attract the financing required for development and pave a path to adoption by providers and payers, all of which are essential for your drug to reach patients.

Embedded in commercial assessment is the clinical path to market approval. Understanding how choice of clinical indication and patient selection strategies can mitigate risk and impact development costs and timeline are important considerations in this assessment. Fortunately, investors and companies do not expect detailed market values from you, the inventor. They will appreciate whether you understand how your product will fit into the competitive landscape and how this impacts its ultimate value and the path to success.

The Bottom Line
Before investing in a novel therapeutic, investors (e.g., a company or venture firm) will want to ensure that future revenues will justify the development costs and provide a positive return on investment. When analyzing potential future revenues, it is important to consider the number of eligible patients (total market), your product's market share, and price. Pricing and market share will depend upon the unique advantages that your product offers over existing therapies and other potential new competitors.

5.9   Making a Compelling Pitch to Potential Investors

Leon Chen

Effectively communicating your pitch to potential investors is a critical skill in attracting the necessary resources for starting a company. While investors may vary significantly in their criteria for evaluating new opportunities, common themes and practices can help entrepreneurs better communicate their ideas. This section reviews both substance and style of a successful pitch.

Box 5.14: The Basics of Every Pitch
1. What problem are you trying to solve?
2. What is your solution to the problem?
3. How do you know the solution will work?
4. Who else is trying to solve this problem?
5. How big is the opportunity?
6. What will it take to achieve success?

5.9.1 What Is the Problem?

There are fundamental questions that every pitch should address. First, what is the problem you are trying to solve? Second, perhaps less obvious but equally important, who cares? An understanding of which technical improvements in efficacy, safety, or convenience will result in a significant commercial opportunity can be a more challenging question to answer. Unlike other industries where a new product is not objectively better, but just different (food, music, etc.) the pharmaceutical industry and the gatekeepers to market access, such as health insurers, require that new products be objectively better than standard of care. Although there are a number of ways to marginally improve upon existing options, market adoption from payers and patients is more likely to be driven by safety and efficacy. Therefore, pitches that solve for these two parameters will generally be more appealing.

5.9.2 What Is the Solution to the Problem?

Moreover, is your solution commercially practical and viable? To realize their full potential, technically elegant solutions to difficult problems must also be convenient and fit easily within the current commercial environment. A common trap for scientist entrepreneurs is to develop solutions that can yield a significant safety or efficacy benefit, but at the cost of convenience or the right physician or payer channels to support the market. For example, the oncolytic virus tamilogene laherparepvec (Imlygic) showed profound efficacy data in melanoma, but has limited market adoption because it requires direct tumor injections, a practice that few oncologists are comfortable doing. Similarly, very high-priced therapies with a high cost of goods (gene and cell therapies) have limited market adoption. A discussion of the pros and cons of your innovation from the point of view of patients, prescribers, and payers can help clarify the likelihood of clinical and market acceptance. You must convince potential investors that your solution is both feasible and attractive to those who will drive its adoption.

5.9.3   How Do You Know That the Solution Will Work?

Making the case that you have a practical solution to a significant unmet clinical need is a combination of strong biology, evidence that a molecule that can produce the desired activity, and a clinical path to proving these effects in patients. Strength of the biology is the underpinning of any proposed solution, and investors will question the rigor of the science and more importantly the hypothesis that one can drug a target or pathway in a safe and effective manner. Early startups will not necessarily have evidence of a molecule in hand at the time of a financing. However, it is important to create a rational plan to discover a molecule with the desired properties. It is important to consider the pharmacological approach here since route of administration, dosing frequency, and cost of goods may limit patient access. Lastly, a thoughtful clinical path to proving safety and efficacy is necessary even at the earliest stages of pitching a company and program. Not all significant clinical unmet needs have obvious clinical strategies that would be tractable for a small startup. Thinking through unique de-risking approaches to clinical development will help sell a program and company.

5.9.4   Who Else Is Trying to Solve This Problem and How Big Is the Opportunity?

Dedicating a few slides to the size of the commercial opportunity, including competitive landscape, is always useful in a pitch. This section of a pitch can not only define the impact of your proposed solution, but can be used to differentiate your approach from others in the field. To determine the market size, refer to Sect. 5.7. In the context of the pitch, if the target market segment is well understood to be large and underserved, do not spend significant time going through the market. If the target market is a niche segment that is more complicated to explain, it is valuable to help investors understand the size of the opportunity.

Reviewing the competition is typically a more valuable exercise. Compare your solution not only to existing products, but also to those in development. It is important to factor in how the competition might impact the standard of care when describing what the market may look like when your solution is approved. Describing the advantages of your approach over current standard of care and new treatments in the pipeline will help attract the required investments.

5.9.5   What Will It Take for This to Be a Success?

A successful presentation will lay out a high-level vision of a business plan and path forward. For product-based companies, an overview of the necessary steps and time to commercialization should be included in your pitch. Technology-based companies should focus on what would be necessary to enable a platform technology to the next meaningful step. Diagnostic companies should be capable of outlining plans into commercialization and what the requirements would be to generate revenues and get to cash-flow break even. Importantly, innovators should have a general sense of financing needs to get to the next value inflection point as well as cash needs to reach overall success; the near-term financing needs to form the basis of the "ask" in any pitch.

5.9.6   The Style

These above points cover the basic content of any successful pitch. The next section will focus on presentation and how you can best prepare for a successful delivery.

Unlike the typical audience for an academic talk, investors vary widely in their background, reflecting a broader range of technical, medical, and business experience. When preparing for an effective pitch, take the time to learn about the investors' areas of expertise and the kind of companies in which they invest. Tailor your pitch to your audience, adjust the depth of scientific detail, set the speed and pace of delivery, and anticipate likely questions based upon the investors' background. The goal of a first presentation is to provide a concise, yet informative overview of your business concept and to get to a second, more in-depth discussion. As a general rule, the presentation should not be overly lengthy and should get to the impact slide quickly to avoid losing the investors' attention. The key challenge is including all the relevant information in the presentation while maintaining efficiency.

> **Box 5.15: Thoughts About Pitch Style**
> 1. Know your audience.
> 2. The pitch is an engaging discussion.
> 3. Highlight challenges and limitations.
> 4. Prepare to pitch by pitching.

5.9.7   Prepare to Be Interrupted

Scientific presentations are typically structured as a complete presentation followed by questions and answers. In contrast, a successful business pitch should be an engaging conversation. You should expect and hope for a back-and-forth discussion covering a range of questions that can take the presentation in any direction. While this can be disruptive and disconcerting when giving a presentation, successfully navigating this type of presentation is often an indicator of a well-communicated story. The ability of an innovator to engage in conversation covering a diverse set of topics beyond the science itself is a very positive sign.

5.9.8   Point Out the Limitations

No business plan is perfect. The most compelling business pitches do an excellent job of identifying and addressing the most significant challenges and limitations. In many cases, the most challenging aspect of a plan cannot be solved technically. The solution may be sought in different business strategies, regulatory tactics, and/or future research to identify new approaches. Avoiding a discussion of these hurdles delays the inevitable and limits opportunities to gain valuable feedback on how to solve these problems. Addressing them proactively will contribute to an engaging discussion and demonstrate a well thought out development plan.

5.9.9 Practice, Practice, Practice

With so much to cover and the uncertainty of what will stick with potential investors, what is the best way to prepare for a successful pitch? The most effective way to refine your presentation is to practice your pitch with other industry veterans. People with varying industry backgrounds will help identify questions and challenges that are likely to arise. In these risk-free discussions, you can learn how to answer the questions effectively, as well as strategies to address potential challenges an innovator might face.

> **Box 5.16: What Surprised an Academic?**
> When looking for funding to start KAI, we talked to hundreds of people: from industry friends, to academics, entrepreneurs, and many, many investors. We learned a great deal from these presentations, and—as expected—heard endless numbers of 'nos'. The key was to treat each new pitch as the one that would secure us the funding. Optimism, determination, and being really well prepared eventually got us what we needed.—*DM-R*

> **Box 5.17: Ask the Experts**
> Although academic researchers are often not familiar with many of the issues related to drug development, take the time and learn them. There is ample literature as well as information on the web. Consult practicing physicians in the field as well as regulatory scientists and clinical trial experts. Talk to as many of them as you can and learn who is also a key opinion leader (KOL); consider recruiting KOLs to your advisory board.

5.9.10 Drug and Diagnostic Pitches

Apart from the general commentary on what makes a successful pitch, there are some specific considerations that a therapeutic drug or diagnostic business plan should address. Businesses focused on developing a novel therapeutic should focus more on the biology the drug is interrogating, the molecular approach to pharmacological intervention, and the clinical path to success. Clinical development elements include a biomarker strategy, pharmacokinetic and pharmacodynamic signals for early signs of efficacy, clinical signal to noise, patient selection for enrollment, and the regulatory path to approval. While some therapeutic areas have fairly straightforward clinical paths, others will require a thoughtful and creative strategy to mitigate financing risks. All drug development plans should address these therapeutic strategy issues.

Diagnostic innovations (see Sect. 2.11) face very different hurdles and thus should focus on different challenges. Demonstrating the technical feasibility of a diagnostic test with retrospective data is a typical early sign of enthusiasm. However, studies done with prospective samples and performed with real operators to create "receiver operating characteristics" or ROC curves are more convincing. While drug companies need to navigate a path to FDA approval, diagnostic innovators need to think carefully about adoption from physicians and payers. Innovators need to consider the impact of their solution on how physicians will manage their patients. The best solutions fit naturally into current patient flow through the healthcare system and impact the therapeutic approach, resulting in improved patient outcomes and reduced overall costs of healthcare delivery. Clinical adoption can be further enhanced if the diagnostic improves revenue for the treating physicians.

Payers have specific mechanisms for payment of diagnostic tests. Some tests can be reimbursed as part of capitated systems where they are included in overall spending of patient care. Others can be paid for through stacked coding of existing codes. While this can be a path to early revenue, these paths are often not sustainable to achieving very large growing revenues without the test requiring its own code. Common themes in these payer dynamics include demonstration of improved clinical utility and a net cost savings to the healthcare system. Clinical utility is defined as not only a better means of diagnosing and separating patients for different paths of clinical care, but the new treatment solution for these patients also results in an improved clinical outcome. The test will also need to show that even with the additional cost of the test being offered, there will be savings elsewhere in the system to offset these new costs. The best diagnostic pitches from an innovator will be very thoughtful about not only how their solution works, but how they plan on proving these critical elements to adoption.

Box 5.18: Common Challenges
1. Common drug development challenges include clinical signal to noise, patient selection, and regulatory strategy.
2. Common challenges for diagnostic companies include changing physician behavior to adopt a novel test and obtaining reimbursement from payers.

The Bottom Line
A compelling pitch to potential investors covers a broad set of topics while communicating the message in a concise manner. This section briefly covers many of the topics that should be included in your pitch. Entrepreneurs should be prepared for an engaging discussion with investors that can lead in a variety of directions. The best way to prepare for a presentation is to pitch to industry experts with a diverse set of backgrounds. With each interaction, the entrepreneur can gradually refine and improve the presentation to better suit the next investor audience while also preparing for any direction the pitch may take.

5.10   Venture Capital Funding

Nina Kjellson

From the inception of the biotechnology industry in the 1970s, and its acceleration over the subsequent decade at Cetus, Genentech, and Amgen, there has been an evolving and symbiotic relationship between venture (adventure?) capitalists and scientific entrepreneurs. These longstanding collaborators marry visionary technical concepts with high-risk dollars in the pursuit of entirely new categories of medical interventions. Without significant "all-or-nothing" financial investment, programs like beta-interferon, recombinant IL-2, PCR diagnostics, and, more importantly, our entire industry would never have seen the light of day.

Even today, as we see a ten-year-plus life science bull-market pull back dramatically, with the initial public offering (IPO) window all but slammed shut after two record-breaking years of new issuances—academic and industrial innovators can rest somewhat easy in the knowledge that healthcare venture firms raised more money last year than in the history of healthcare venture capital. Investors from private and public pension plans and endowments, family offices, and fund-of-fund institutions poured over $40 billion into 244 discrete funds that sit poised to deploy capital into private companies (PitchBook 2023).

Box 5.19: What Is Venture Capital?

From Investopedia:

Venture capital (VC) is a form of private equity and a type of financing that investors provide to startup companies and small businesses that are believed to have long-term growth potential. Venture capital generally comes from well-off investors, investment banks, and any other financial institutions. However, it does not always take a monetary form; it can also be provided in the form of technical or managerial expertise. Venture capital is typically allocated to small companies with exceptional growth potential, or to companies that have grown quickly and appear poised to continue to expand.

Though it can be risky for investors who put up funds, the potential for above-average returns is an attractive payoff. For new companies or ventures that have a limited operating history (under two years), venture capital is increasingly becoming a popular—even essential—source for raising money, especially if they lack access to capital markets, bank loans, or other debt instruments. The main downside is that the investors usually get equity in the company, and, thus, a say in company decisions (Hayes 2022).

Historical Context

When I began my career as an investor in 2000, there were only 106 such funds, deploying $19.59 billion. Tech was in its heyday and biotech an ugly duckling, floating on a rising tide. When the tech bubble burst, a government ban on patenting human genes made the sequencing of the human genome a great feat for science but lackluster for business, and biotech joined the pall falling over the general markets. VC purse strings closed nearly as fast as the IPO window did and took quite some time to recover. In 2006, when I was a proud mentor to the launch of the SPARK program, 100 healthcare VC funds raised $16.75 billion dollars. How far we have come (PitchBook 2023).

Not only has the venture industry scaled significantly, but the model for venture has also expanded with time, and the environment has proliferated with a range of fund and investing models. This venture model maturation reflects a growing appreciation of the value the life science industry brings to the economy at large: job creation and economic output, public health stewardship, caretaking of an aging population, and wealth creation via technology translation. The COVID-19 pandemic also drove a surge in VC activity with attention on biotech's contribution to diagnostic testing, vaccines, and antiviral drugs, and the digital transformation of healthcare. Investment allocation to diagnostics/tools and to digital health nearly tripled in 2020 and 2021 versus 2019 compared to a doubling for devices and biopharma; proof of the pandemic's fueling (Silicon Valley Bank Financial Group 2022).

With the maturation of the industry comes diversity and specialization of venture capitalists (see Table 5.1 for various stages and types of biopharma investing). The template for success for an entrepreneur seeking capital is no longer one size fits all or even one size fits many. A founder is wise to do their research before going to market in search of venture investment and to ask themselves focused questions about what they are looking for: mere investment? Strategic advice? Network of advisors and contacts? Investors are increasingly specialized, not only by size and investment stage as described above, but also by subsector (biopharma, digital health, healthcare services, medical devices, diagnostics, life science tools, etc.) and even subspecialized within therapeutic areas or applications (e.g., oncology, longevity, women's health, cell or gene therapy platforms). When looking for VC investment, it is critical to know your target audience. In today's age of social media and personal and professional branding, there is no reason you cannot also get a good sense for an investor's preferences and biases—as well as unique value-add before you meet. When possible, a personal introduction versus a cold call can provide a significant advantage. Come prepared with what you want and when possible, tailor your pitch according to the interest of the VC.

Table 5.1 Stages and types of biopharma investing

Stage	Type of activities	Relationship
Incubator/ accelerator	Form companies; typically, solo-investor	Provide operational support
Early-stage	Seed to series A/B	Actively involved
Mid-stage	Series B to last private round	Moderately active
Crossover	Last private and IPO/public	Typically, more passive
Stage-agnostic	All rounds +/− IPO and public	
Hedge fund	Long and short investing	Typically, more passive
Long-only mutual fund	Public investing with no hedging	Passive
Private equity/ buy-out	Controlling stake. Equity investor +/−debt (leveraged buy-out).	Active or even activist; seeking board control to drive changes
Venture debt	Lender +/− equity or warrants	Passive
Corporate venture capital/strategic	Venture capital fund of pharma, MedTech etc.	Involvement varies. Strategy typically aligns with parent company vs financial returns.

Abbreviations: *IPO* Initial Public Offering

Box 5.20: What Surprised an Academic?
When searching for funds, remember that venture capitalists can bring a great deal more to the table than their firm's financial support. For example, KAI's initial investor, Skyline Ventures, provided critical mentorship to the founding team, offered rent-free working space in its offices, helped assemble a cross-functional team of consultants for our virtual company, and played a leading role in securing additional investors for the series A financing. While money is indispensable when starting a company, also consider the additional support that a firm offers when selecting your VCs.—*DM-R*

5.10.1 Opportunity Assessment: A Template

Similarly, when meeting with entrepreneurs, we have a set of key questions we tend to ask:

- Is the innovation sufficiently meaningful to be viable? Is the unmet need life threatening or highly burdensome? Is there high empathy for patient, provider and the healthcare system?
- Is the novelty to risk ratio acceptable? Pick one, maybe two major risks: biology, technology, manufacturing, market, etc.
- Is the team world-class? If the founders lack experience or expertise, do they have access to people who can provide it?
- Can the idea translate to outcome in an acceptable timeframe? It will always take longer and cost more than planned. Factoring this in, is there positive return on investment?
- Is there defensibility? Is there IP (patents or know-how) or some other competitive edge?
- What is the full capital requirement to the Promised Land? Even if there is an exit inflection point for founders and VCs far sooner than commercialization, you

must understand the total cost to market because it will be calculated into the valuation at every financing round and at liquidity.

- Does the founder have grit and the X-factor to stand out from a crowd? Entrepreneurship is relentless, fraught with challenges, and ever more competitive. Can this founder and team strategize, manage adversity, communicate transparently, and cultivate a high-performing and resilient culture?

We screen hundreds of opportunities each year. While we are driven by a passion for science, patients, and working with talented entrepreneurs, we are also duty-bound to make the best financial returns for our investors. We model each investment from the bottom up and aim to construct a portfolio to account for the fact that between one third to one half of early-stage investments fail to generate a profit. We benchmark risk and costs for research and development projects against industry metrics and our own experiences. We conduct due diligence to ensure that the medicines we invest in are compelling to treating physicians and to the health plans that will pay for them. We pressure test early-stage profiles with science and business leaders at larger biopharma companies whose pipelines are partly insourced from venture-backed companies and ensure that the data sets produced by startups align with what acquirers need to see to believe. Finally, we marry our investment thesis with the right financing structure. We partner with entrepreneurs to align milestones with increases in business value. This means that we often fund a company for several years through key inflection points, such as preclinical and clinical data generation, advancing money in discrete tranches and partnering with a syndicate of like-minded investors to carry the company long term.

When companies have more than one program, we often explore how strategic partnering might underwrite the pipeline as compared with using equity dollars. As much as fresh money lies ready to deploy in venture funds, there is also a lot of stored-up dry powder on balance sheets at large biopharmas, eager to generate more drug pipelines. For entrepreneurs and investors alike, partnering is a less dilutive way to build shareholder value. This is where the capitalism part of venture capital comes around again: To reduce losses and maximize overall investment returns, startups (and investors) need to focus as well as minimize dollars exposed at the riskiest time points.

Whether we are in for a correction or merely a calibration of the markets, two things are true: We are blessed with an embarrassment of riches in two dimensions; scientific innovation, and available private capital. These are opportunities not to be squandered with incredible opportunities for patients and profits when the two are combined.

Box 5.21: What Surprised an Academic?

It is easy to think of an entrepreneur:investor relationship as very transactional. The one has an idea and the other, money. In order to get the cash, the founder has to give up control and make a 'deal with the devil.' The earlier the stage, the more ownership has to be yielded. In reality, however, the relationship can be a true partnership. Drug development especially tends to be a long and harrowing journey, but with a very virtuous goal: helping patients. If both sides keep that vision front of mind, it can be a wonderful and reciprocal enterprise.—*NSBK*

Box 5.22: The Life Science VC Value Equation
When an innovation is novel and an experienced - or especially talented - team can demonstrate meaningful impact via measurable outcomes in a rational timeframe on a reasonable budget and generate an *outsized return*.

5.11   Legal Aspects of a Startup Biotechnology Company

Updated by: Haim Zaltzman and J. Jekkie Kim
Original authors: Alan Mendelson, Peter Boyd, and Christopher M. Reilly

The first question that academic entrepreneurs affiliated with universities or other research institutions must ask when thinking about starting a biotechnology company is whether it is the right time to form a business entity, or continue to grow and expand research and development activities within the comfort of the research institution. The decision should, in part, be driven by a cost–benefit analysis—that is, do the benefits of forming a business entity outweigh the initial and ongoing costs of maintaining it? One of the benefits of a business entity is that, if properly structured and capitalized, it provides limited liability—shielding officers, directors, and investors from personal liability for the liabilities incurred by such business entity.

For academic entrepreneurs, there may be additional, unique considerations in deciding whether or when to form a business entity. Academic entrepreneurs have access to lab space, administrative personnel, core service centers, library facilities, grant opportunities, and related infrastructure that allow them to perform research and development activities at a fraction of the cost of their industry competitors. Furthermore, staying within an institution may provide a more reliable source of income.

On the other hand, remaining within an institution generates additional risks above and beyond the generic costs associated with forming a business entity. For example, in the US, an academic entrepreneur must assign any and all of their intellectual property (IP) rights conceived or reduced to practice as an employee to their institution. This means that academic entrepreneurs lack control over the destiny of any IP they have invented within the institution, including monetization strategies, unless and until the academic out-licenses such IP rights into separate business entities from the applicable institutions.

In sum, if the entrepreneurs are still investigating academic theories that lack commercial value, they may wish to remain within the institution. This way, they continue to leverage the advantages of the institution until they determine that there is a viable business opportunity that justifies out-licensing the IP rights to a business entity. If an academic entrepreneur identifies potential business opportunities to develop and commercialize products, services, or other technologies that are based on IP rights they generate, then the academic must weigh the benefits, costs, and the timing of out-licensing such IP rights from the respective institution into a business entity.

Box 5.23: What Surprised an Academic?
In the United States, university and research institutions contractually own any and all IP that are created by an academic. Even when an academic is listed as the "inventor" of a patent application, the patent application is then "assigned" to the institution, i.e., the institution owns the patent application and has the right to exploit the IP without any additional consent from the academic. Therefore, for an academic to commercialize his/her invention, the IP rights must be out-licensed to a separate business entity.

Forming a startup company requires experienced corporate attorneys very early in the process and before signing any documents. Experienced attorneys will protect the interests of the business and provide guidance to the founders on ownership issues and what they should expect to retain after angel or VC funding, employee hiring, and other essential processes. It is prudent to retain your own lawyer to represent your personal best interests, rather than to rely on university, VC, or company lawyers.

Resources are available to help a startup generate basic incorporation and founder documents (see Sect. 5.11 Resources).

Once entrepreneurs determine that it is time to form a business, they must select the best type of entity for their business. Business entities fall into two broad categories: those that are subject to "pass through" taxation and those that are not. In pass through entities, the profits and losses of the company are passed through the company to the investors and only taxed once. Pass through entities include partnerships, Limited Liability Companies (LLCs) and Subchapter S Corporations (S-Corps). The more common Subchapter C Corporation (C-Corp) is a non-pass through entity, where income or losses to the corporation and dividends distributed to shareholders are taxed separately. When selecting the appropriate entity, entrepreneurs should assess all options based on what will work effectively, without creating too much burden on a limited administrative staff, and whether the platform chosen is flexible enough to meet the short-term and long-term goals and challenges of a growing biotechnology company.

Biotechnology companies are capital intensive relative to most other industries and take an extended period of time to generate revenues, so entrepreneurs need to ensure that all possible funding sources remain available. Two common funding sources at the outset are angel investors and venture capital firms (VCs). VCs will rarely invest in pass through entities because they create negative tax consequences for their limited partners, many of which are tax-exempt institutions. If an entity plans to approach VCs, it will most often need to be structured as a C-Corp. Occasionally, a few wealthy individuals or angel investors are prepared to fund the early research and want pass through taxation so that they can promptly write off their investment. It is extremely unlikely, however, that these angel investors will be able to fund the company beyond its early research phase, so at some point, the entity is typically converted to a C-Corp to raise capital from institutional investors.

Partnerships and LLCs require relatively complex governing agreements that tend to be time-consuming and expensive to create and explain to investors and early employees. These additional costs associated with forming and operating an LLC or partnership combined with pass through taxation make them inappropriate for most biotechnology startups. Whereas it is possible to form an S-Corp for pass through taxation and convert it to a C-Corp when the company looks to raise venture capital, most investors recognize and will insist that the long-term benefits of the rights, preferences, and privileges of preferred stock permitted under Subchapter C outweigh the short-term benefit of pass through taxation. In our experience, most biotechnology companies decide to incorporate as a C-Corp once they have considered the flexibility and benefits that a C-Corp can provide.

Assuming that the entrepreneurs select a C-Corp as their desired entity, they will next need to select a state of incorporation. Most institutional investors in the United States are used to investing in Delaware corporations. Delaware also has a well-developed body of corporate law and efficiently deals with corporate issues at both the administrative and judicial levels. However, choosing to incorporate in Delaware will require the company to qualify to do business in its state of operation and to incur a second, relatively small franchise tax. On balance, most biotechnology companies are formed as Delaware C-Corps. Some biotechnology companies choose other states, such as California, Nevada, or New York to incorporate, but Delaware remains the dominant state for incorporation.

Recruiting and maintaining the best team is a key component of success for every biotechnology startup. Therefore, entrepreneurs want to ensure that they can effectively attract and compensate talent. At the outset, entrepreneurs want to determine who will be a founder. The concept of who is or who is not a founder can assume a significance not warranted by any legal recognition or long-term consequence. However, being identified as a founder often provides an entrepreneur with a certain leadership benefit. In addition, founders typically assume a greater level of risk than employees, consultants, or advisors who join the company after funding has been obtained. Thus, the founders are normally granted the opportunity to purchase equity at a nominal cost when the corporation is initially formed, recognizing their contribution to the basic concept or idea.

Whether stock is sold just to the founders or to a broader group of employees, consultants, and advisors, it is important to document the initial allocation of equity to mitigate the risk of future disputes among contributors to the enterprise and to ensure that all stock and option issuances are made in compliance with securities laws. In this regard, entrepreneurs must tackle the fundamental decision of whether to require vesting of the stock or options to be issued—for themselves and future employees, consultants, and advisors—and what type of employment or consulting agreement should govern the various individuals' service relationships with the company. Under a vesting scheme, founders and employees are allowed to purchase a block of common stock, but only gradually gain full ownership of that stock on the timeframe described in the vesting schedule. Vesting is quite common and is more protective of the entrepreneurs relative to each other and investors, but stock subject to vesting will likely require a filing under Section 83 of the Internal Revenue Code and more time and expense with counsel.

In any event, all parties to the enterprise should make sure that every contributor is party to some form of proprietary invention and assignment agreement that grants IP rights to the company. In evaluating these issues, entrepreneurs need to be mindful that potential investors are unlikely to invest if they do not have comfort that key personnel are committed for the long term and that the company's IP is being protected. Protecting IP is key for biotechs!

After the above core issues have been addressed, entrepreneurs need to consider the size and structure of the company's board of directors, as well as any special governance provisions that ought to be included in the company's charter or bylaws. An experienced emerging company corporate lawyer can guide entrepreneurs through all of these issues and help the company document key decisions, develop a set of standard documents to be signed by employees and consultants, and obtain a tax ID (an employer identification number) so that the company may open a bank account and pay employees and make all state and federal regulatory filings associated with the sale of securities in a timely manner.

The Bottom Line
Biotechnology startups face a unique set of challenges when growing from a few founders performing research to a fully funded company advancing product candidates to clinical development and commercialization. Like the start of any significant business, it is essential that entrepreneurs take the proper first steps as they pursue an opportunity from idea to business entity and beyond.

Founders should remember that research institution or company lawyers are not their lawyers. In all legal aspects, founders should engage their own lawyers for legal advice and protection of their interests.

Resources
- Resources that allow a startup to generate the basic incorporation and founder documents: https://www.lathamdrive.com/resources/documents

5.12 Plan, Organize, Motivate, and Control

John Walker

With your lead program, capital, and investors in hand, you might be tempted to believe that your work is done. After all, you have come a long way since your initial discovery in the lab or clinic. But the realization soon hits that you now have to execute on that pitch you made to the venture investors. This section describes how to lay the foundation for building a successful company.

In simple terms, this requires understanding some basic principles of running a business; namely, the need to plan what you intend to do, organize your people and resources to accomplish the plan, motivate your team to accomplish tasks in a timely manner, and finally, implement a control mechanism to keep your company on course.

> **Box 5.24: What Surprised an Academic?**
> Sometimes academics make the incorrect assumption that, because companies are engaged in for-profit drug development, industry professionals are chiefly motivated by money. In fact, many in industry are motivated by being part of the process to alleviate human suffering. Many are idealists!—*DM-R*

5.12.1 Plan Your Course

The planning process can be broken down into several key elements. The first is a clear and understandable mission statement—a short paragraph or sentence that reflects the long-term vision of your new company. For example: *To develop and commercialize new treatments for solid tumors based on the modulation of kinases.* When developing a mission statement, consider the company you intend to have 10 years from now, not what you have today or will have next year. The mission needs to be aspirational, and each word should be carefully selected to convey the vision clearly and concisely.

Next, closely following the mission statement, articulate a value statement for your company. The value statement will guide development of the company culture that will best help you accomplish your goals. For example: *We will conduct our research in a rigorous, peer-reviewed process; we will achieve our business objectives in an open and transparent manner; we will recognize the value of each of our associates and encourage diversity in thought and experience.* The main point is to think about the type of company you want to be and state that as clearly as possible.

Now that you know the long-term vision of the company, as well as the type of company that you want to create, the next step is to outline just how you intend to get there. This requires a multiyear strategic plan, one that should look ahead at least five years. Because the biotechnology and biopharma industries, like many others, are volatile and changing rapidly, this strategic 5-year plan should be reviewed and revised on an annual basis. In developing a strategic plan, it is critical to be open and honest about your capabilities (strengths) and areas for improvement (weaknesses). One approach is called an environmental analysis, also referred to as a Strengths, Weaknesses, Opportunities, and Threats (SWOT) analysis (Fig. 5.2).

A comprehensive and honest assessment of your strengths and weaknesses will help determine which elements of your business need to be preserved, strengthened, or changed. Ideally, this exercise will allow you to focus on the most critical issues for success, including intellectual properties, capital, investor base, management team, scientific capabilities, board of directors, employees, skills, scientific advisory board, etc. Make it clear that this analysis is for internal use. Importantly, in going through this exercise, recognize that there are no "sacred cows"; everything about the company should be analyzed and debated as either a strength or a development need.

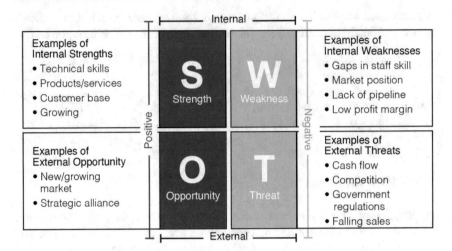

Fig. 5.2 Strengths, weaknesses, opportunities, and threats (SWOT) analysis chart

Once you have reviewed your strengths and weaknesses, assess the opportunities and threats your company faces. This requires an examination of trends in the market, conditions for raising new capital, prospects for company partnerships, competitors, FDA or other regulatory agency issues, changes in clinical practice, and changes in the health insurance or payer landscape. Review all of the external issues that may have a bearing on your company's success.

> **Box 5.25: What Surprised an Academic?**
> I initially felt that this "kumbaya" exercise was somewhat embarrassing and that the company's mission, culture, and long-term goals should be obvious. Through working on these with the team, I realized that they were not clear even to me.—*DM-R*

Consider the following statement of a threat: *FDA is increasingly concerned with safety, narrowing the severity and types of adverse events allowed in treating cancers*. Obviously, this observation will influence your ability to reach your eventual goal of commercializing new treatments for solid tumors, since FDA approval is necessary to introduce a product to the market. Explicitly identifying this as a threat provides an early guidepost to plan your clinical development activities.

With a mission, values statement, and environmental analysis in hand, you are now ready to start reviewing the goals and strategies that will help you achieve your desired outcome. This part of the planning process should incorporate five to eight overall goals that can be accomplished within the 5-year horizon of the plan and the strategies that will need to be employed to reach each one.

An example would look as follows:

Goal	By the end of the planning horizon, we will have completed a phase 2b study on our lead program demonstrating an increase in survival of more than 6 months in pancreatic cancer.
Strategies	1. Advance two lead compounds to toxicity studies by the end of the first 6 months of operation.
	2. Nominate one of these compounds for clinical development by year end.
	3. Complete a phase 1/2 study in 25 subjects at four centers by the end of the second full year of operation.
	4. Complete enrollment of 75 patients at eight different centers in the US and Europe as part of the phase 2b study by the end of year 3.

In developing your goals, remember to touch on all of the key aspects of your company. These should include product and clinical development, capital formation strategies, personnel policies and philosophies, business development goals, etc.

You will find that a good strategic plan will flow naturally into an annual operating plan (budget) for the next year, allow you to set individual Management by Objective (MBO) goals for all of your key people, align the work of the team, provide measurable milestones for determining your progress, and allow you to more successfully articulate your company's vision and goals to outside parties such as prospective investors, partners, and employees.

5.12.2 Organize Your Resources

With the plan now in hand, you can put together your company organization. This does not just refer to an organization (Org) chart, but rather the systems you will need to put in place to realize your goals. The members of your Board of Directors, Scientific Advisory Board, Clinical Advisory Board, executive team, and scientific staff must possess the requisite expertise to accomplish your plan. As an example, if one of your goals is to raise a new round of financing in year 2, hiring a Chief Financial Officer may be an important strategy to articulate within the first year of operation; if partnerships are important, then a Business Development function may be critical. The timing and need for a Clinical Advisory Board will also be more apparent once the plan is in place. Use the plan as your guide to develop your organization at all levels, recognizing the fluid nature of the skill sets and expertise that you will need at different time points in your plan.

5.12.3 Motivate Your Team

Highly motivated personnel are critical to accomplishing the company goals in the desired timeframe. Appropriate incentives can help maintain the desired corporate culture and ensure that individual and collective interests are aligned. It is important to recognize that most people are not motivated by money, which is more of a "hygiene factor," but rather by such things as the scientific merit of the project they are working on, their relationships with coworkers, and the belief that the company values them both as individuals and for the contributions they make to the organization.

While many of these factors may seem intangible, they are critically linked to your company's culture, management practices, and personnel policies. These should be specifically addressed in your planning process. For example, if you state that you want to have a performance-based culture, then are you putting in place those types of incentives that reward and recognize success? Do you have hiring practices that allow for interviews that determine "cultural fit" to the organization? Does your practice call for interviews on a 360° basis (i.e., do potential direct reports have the chance to interview their prospective new boss)? Do you have a promotion-from-within philosophy, where individuals can advance their careers as opportunities present themselves within your organization? Do you want to establish a bonus program at selected levels or all levels in your company? Is the bonus based on clearly stated and quantifiable goals? Do you practice open and honest communication with all of your associates? An easy way to accomplish this is to hold monthly "all hands meetings," where many or all aspects of the company are presented and discussed with all employees. The combination of promising scientific programs with an open, honest culture that values individuals will lead to employee retention, satisfaction, and team building.

Box 5.26: What Surprised an Academic?

We all know about teamwork; after all our labs are teams of scientists. However, in a company, we truly have a single goal—the success of the company. The difference is apparent from day one of the company, when you realize how easy it is to derail the company by missing a milestone, for example, because one person in the team did not do their job. And when you walk out of the first board of directors meeting with the investors to see the expectant faces of all the employees, you also realize that such teamwork is truly exhilarating.—*DM-R*

5.12.4 Control Your Progress

Once you are up and running, the last item to consider is that of control. Perhaps this is more easily understood as "knowing if you are still on track." The first step to ensuring that you keep on track is to monitor and measure achievement in real time. A common mechanism for tracking progress is to compare budgeted expenses versus actual expenses. Perhaps equally important, management should monitor progress against stated objectives and timelines on a routine basis. Furthermore, if you have a bonus program for all employees, then reporting progress on a regular basis (perhaps at those monthly all hands meetings) is important. If important goals are not being accomplished in a timely fashion, it is critical to determine the cause, refocus your efforts, and take corrective actions to ensure the on-time delivery of your objectives.

Box 5.27: What Surprised an Academic?

When KAI decided to terminate one of the clinical programs, I attended the failure party that John Walker organized—despite my doubts on the appropriateness of celebrating such an event. The management emphasized the excellent work conducted by the team to enable "failing fast." By showing lack of a path forward, the team freed the clinical group and budget to focus on another indication earlier, the one that resulted eventually in Amgen acquisition of KAI. The failure party was also an opportunity to recognize the team's hard work on failure analysis, which could help the company succeed the next time.—*DM-R*

The Bottom Line

Plan what it is you intend to do.
Organize your people and resources to accomplish the plan.
Provide *motivation* for accomplishing the plan.
Provide *control* mechanisms to keep you on course.

Key Terms and Abbreviations

Key Terms

360° Evaluation: Evaluation of an employee (or future employee) by supervisors, peers, and those who report to that individual.

All Hands Meeting: A periodic company meeting that provides an update to all employees. This is a powerful and simple tool to motivate and control.

Best-in-Class: A drug that offers a clinical advantage over what is currently available on the market for a given indication.

Board of Directors: The group of individuals representing investors/owners who set the strategic direction and have ultimate decision-making authority for the company.

Code of Federal Regulations (CFR): Rules and regulations published in the federal register by the executive departments and agencies of the US Federal Government.

Conflict of Interest (COI): An individual's private interests or considerations of personal financial gain that could compromise that individual's professional obligations.

First-in-Class: A drug that uses a novel and unique mechanism of action to treat a medical condition.

Freedom to Operate (FTO): The ability to use or commercialize a product or process without infringing another party's valid intellectual property (IP) rights.

Human Research Protection Program (HRPP): A collection of policies, guidances, and supporting documents governing human subject research and the protection of participants.

Incidence: The number of people in a population who develop a disease or other health outcome over a period of time.

Initial Public Offering (IPO): Offering shares of a private corporation to the public in a new stock issuance for the first time.

Institutional Review Board (IRB): A committee formally designated by an institution to review, approve the initiation of, and conduct periodic reviews of biomedical research involving human subjects.

Intellectual Property (IP): A category of property that includes products or inventions of original thought with legal protections (such as patents) against unauthorized use by others.

Investigational New Drug Application (IND): Document filed with FDA prior to initiating research on human subjects using any drug that has not been previously approved for the proposed clinical indication, dosing regimen, or patient population.

License: A legal agreement granting another party rights to use intellectual property.

Licensee: A third party that pays for access to technology.

Limited Liability Company (LLC): A business entity that protects partners or shareholders from being liable for the company's financial losses and debts. Liability is assumed by the business rather than the partners or owners. LLCs are subject to "pass through" taxation, that is, the profits and losses of the company are passed through the company to the investors and only taxed once.

Management by Objective (MBO): Strategic management model to improve company performance by clearly defining objectives that are agreed to by both management and employees.

Neglected Tropical Disease (NTD): A diverse group of 20 conditions that are mainly prevalent in tropical areas, where they mostly affect impoverished communities.

New Drug Application (NDA): Application to obtain FDA approval for the sales and marketing of a new drug in the US.

New Molecular Entity (NME): A new drug submitted to the FDA Center for Drug Evaluation and Research (CDER).

Not-for-Profit Drug Development (NPDD): Developing a drug for a disease where the costs of development, manufacturing, and distribution exceed any future revenue (e.g., development programs for drugs addressing infectious diseases of low-income countries, for therapeutics for ultra-rare diseases, or for advancing repurposed generic drugs or other therapeutics for which there is no patent protection).

Option: A legal agreement granting temporary rights to intellectual property.

Organization (org) Chart: A graphic representation indicating job functions in the organization and the reporting structure.

Prevalence: The number of people in a population who have a disease or other health outcome at a given point in time.

Priority Review Voucher (PRV): An FDA program that grants a voucher for a future priority review for any drug when a drug developer obtains market approval of a therapeutic for eligible neglected tropical diseases or rare pediatric diseases. FDA's intent is to incentivize development of drugs for these non-profitable indications. PRVs can be sold to other biopharmaceutical companies, commanding prices in the range of $100 million.

Product Development Partnerships (PDP): Organizations that target drug development for economically deprived markets, aiming to develop a product by integrating contributions of diverse partners, and generally funded through philanthropic support.

Research Policy Handbook (RPH): Stanford policies regarding the conduct of research.

Risk Assessment Committee (RAC): A multidisciplinary panel that evaluates studies for atypical financial or administrative risks and makes recommendations regarding clinical research operation and regulatory policy. RAC provides an additional vehicle for risk analysis and mitigation, not overlapping with human subject protections or scientific validity.

Royalty: Payment based on revenue generated from licensed technology.

Strengths, Weaknesses, Opportunities, and Threats (SWOT): Analysis used to assess the company/project and plan accordingly.

Target Product Profile (TPP): A document outlining essential characteristics of the final drug product.

Technology Transfer Office (TTO): The university group responsible for managing intellectual property owned by the university.

US Food and Drug Administration (FDA): The Federal Health and Human Services agency responsible for protecting public health by assuring the safety and efficacy of drugs, biological products, vaccines, medical devices, and other designated products.

Key Abbreviations

APB	Administrative Panel on Biosafety
AMCs	Advanced Market Commitments
ALK	Anaplastic Lymphoma Kinase
ACE	Angiotensin-Converting Enzyme
CCTO	Cancer Clinical Trials Office
CDC	Centers for Disease Control and Prevention
CTRU	Clinical and Translational Research Unit
CRQ	Clinical Research Quality
DSMC	Data and Safety Monitoring Committee
δPKC	Delta Protein Kinase C

DNDi	Drugs for Neglected Diseases Initiative
HIPAA	Health Insurance Portability and Accountability Act of 1996
HREC	Human Research Ethic Committee
IEC	Independent Ethics Committee
INPADOC	International Patent Documents
MDGH	The Medicines Development for Global Health
MMV	Medicines for Malaria Ventures
NIH	National Institutes of Health
NIAID	National Institute for Allergy and Infectious Diseases
OOPD	Office of Orphan Products Development
OWH	OneWorld Health
PCT	Patent Cooperation Treaty
PD-1	Programmed Cell Death 1
PD-L1	Programmed Cell Death Ligand 1
PD-L2	Programmed Cell Death Ligand 2
RCO	Research Compliance Office
RMG	Research Management Group
SRC	Scientific Review Committee
TDR	Special Programme for Research and Training in Tropical Diseases
SIR	Sponsor–Investigator Research
SCRO	Stem Cell Research Oversight
TRND	Therapeutics for Rare and Neglected Diseases
USPTO	US Patent and Trademark Office
WIPO	World Intellectual Property Organization

References

Burnier M, Egan BM (2019) Adherence in hypertension: a review of prevalence, risk factors, impact, and management. Circ Res 124:1124–1140

Emmert-Buck MR (2011) An NIH intramural percubator as a model of academic-industry partnerships: from the beginning of life through the valley of death. J Transl Med 9:54

Hayes A (2022) Venture capital. In: Investopedia. https://www.investopedia.com/terms/v/venture-capital.asp. Accessed 29 Mar 2022

Kim ES, Omura PMC, Lo AW (2017) Accelerating biomedical innovation: a case study of the SPARK program at Stanford University, School of Medicine. Drug Discov Today 22:1064–1068

Mochly-Rosen D, Grimes KV (2011) Myocardial salvage in acute myocardial infarction – challenges in clinical translation. J Mol Cell Cardiol 51:451–453

Moos WH, Kodukula K Nonprofit pharma: solutions to what ails the industry. Curr Med Chem 18:3437–3440

Moran M (2005) A breakthrough in R&D for neglected diseases: new ways to get the drugs we need. PLoS Med 2:e302

Nwaka S, Ramirez B, Brun R, Maes L, Douglas F, Ridley R (2009) Advancing drug innovation for neglected diseases—criteria for Lead progression. PLoS Negl Trop Dis 3:e440

Olliaro P, Seiler J, Kuesel A, Horton J, Clark JN, Don R, Keiser J (2011) Potential drug development candidates for human soil-transmitted helminthiases. PLoS Negl Trop Dis 5:e1138

PitchBook (2023) Retrieved Aug 1, 2023, https://www.pitchbook.com

Schulze U, Ringel M (2013) What matters most in commercial success: first-in-class or best-in-class? Nat Rev Drug Discov 12:419–420

Silicon Valley Bank Financial Group (2022) Healthcare investments and exits. https://www.svb.com/trends-insights/reports/healthcare-investments-and-exits/healthcare-investments-and-exitsannual-2022

Vrijens B, De Geest S, Hughes DA et al (2012) A new taxonomy for describing and defining adherence to medications. Br J Clin Pharmacol 73:691–705

Wouters OJ, McKee M, Luyten J (2020) Estimated research and development investment needed to bring a new medicine to market, 2009–2018. JAMA 323:844–853

Yu H, Boyle TA, Zhou C, Rimm DL, Hirsch FR (2016) PD-L1 expression in lung cancer. J Thorac Oncol 11:964–975

Concluding Thoughts

6

Daria Mochly-Rosen, Kevin Grimes, and Steve Schow

Daria Mochly-Rosen, Ed, Kevin Grimes, Ed.

Advances in drug therapy came fast and furious during the twentieth century, resulting in the cure of many previously fatal diseases such as pneumonia, tuberculosis, and testicular cancer. Treatments for more chronic diseases (e.g., asthma, heart failure, hypertension, and diabetes mellitus) have improved the duration and quality of life for untold numbers of patients. These new therapies arose from basic research discoveries followed by application of the scientific disciplines of drug discovery and development. In recent years, the cost and time required to bring a new drug to market have escalated dramatically, resulting in the consolidation of large pharma companies and a contraction in the basic research component of the industry. In order to maximize return on investment, biopharma companies have increasingly focused on developing treatments that can command premium pricing. Less profitable indications, even critical needs such as new antibiotics for drug resistant bacterial infections, are largely ignored by the industry. This raises two important questions: (1) Where will our next generation of drugs originate? and (2) How we can we address society's need for new therapeutics that may not be sufficiently profitable to gain industry's attention?

We strongly suggest that academic research institutions fill this void by adopting a SPARK-like approach or establishing an applied medicines research and development institute to fuel applied research and engage in early drug discovery, generating drug candidates that are attractive to industry. Academics are well positioned to fill this gap in the discovery pipeline. Free from the pressures of generating a short-term financial return on investment, we can focus on drug development projects that

D. Mochly-Rosen (✉) · K. Grimes
Chemical and Systems Biology, Stanford University School of Medicine, Stanford, CA, US
e-mail: sparkmed@stanford.edu; kgrimes@stanford.edu

S. Schow
SPARK at Stanford Advisor, Stanford, CA, US

address the greatest unmet clinical need. Once projects are appropriately "de-risked," they can be licensed to existing biopharmaceutical or startup companies for further development. Academics can also engage in policy research and advocacy to help establish new financial approaches to support the development of less profitable drugs.

If we agree that academia has a social responsibility to ensure that new drugs reach patients in need, we must be willing to devote more financial and human resources to this applied science endeavor. We will need to develop the facilities and expert staff/faculty in such disciplines as assay development; high-throughput screening; medicinal chemistry; pharmacokinetics and absorption, distribution, metabolism, and excretion (ADME); toxicology; and regulatory science. If we in academia fail to take on this expanded role, we may deprive future generations of novel therapies that will save lives, improve health, and lower the escalating costs of healthcare.

6.1   Addressing the Market Failure for Developing Novel Antibiotics

Steve Schow

Much has changed since the first edition of this volume. The infinite opportunities for creating new medicines have been greatly enriched by the ever-expanding knowledge base of biomedical science, novel approaches and technologies for new drug invention, and the creativity and cleverness of the biomedical researchers engaged in this vast global enterprise. Academic biomedical scientists seem more resolute than ever to transform their frontier curiosity-driven science into practical products for the ultimate benefit of patients with unmet medical needs. Investors' appetite for exploiting the abundance of new biomedical science has grown, as well, to match this wealth of new opportunities.

However, certain major medical needs remain unaddressed by industry, such as the continual necessity for new antibiotics to treat increasing numbers of resistant organisms. The economic models on which the current drug invention and development system were built no longer meet the needs of the global patient population. Before the rise in antibiotic drug use, lethal infections could arise from a simple cut on the hand. The loss of antibiotic therapeutic effectiveness to bacterial resistance now threatens the progress we have made since the 1930s against bacterial infections. The pharmaceutical industry was once the wellspring of antibiotic discovery, yet the industry has abandoned this therapeutic area for more lucrative endeavors.

A 2019 report indicated that an estimated 4.95 million people died globally from illnesses in which bacterial antimicrobial resistance played a role and 1.27 million of those deaths were directly caused by antibiotic resistant organisms (Murray et al. 2022). Despite continuing calls for new antibiotic development, we have failed to develop an arsenal of new agents to control resistant organisms. The much-anticipated resurrection of the antibiotic discovery infrastructure has not materialized.

The lack of private sector investment in innovative drugs to treat serious, neglected, unmet medical needs is a direct result of current pricing, investment return expectations, and high regulatory demands required for modern drugs created by the private, market-driven sector. Before a major crisis occurs, now is the time to seriously discuss the advantages of migrating new drug invention and early-stage development of medicines for neglected therapeutic areas from the private sector into a network of academic, applied biomedical institutes dedicated to the creation and early-stage development of antibacterial and antiparasitic drugs. Challenges and solutions in the development of new antibiotics are illustrated in Fig. 6.1 and are further discussed in this section.

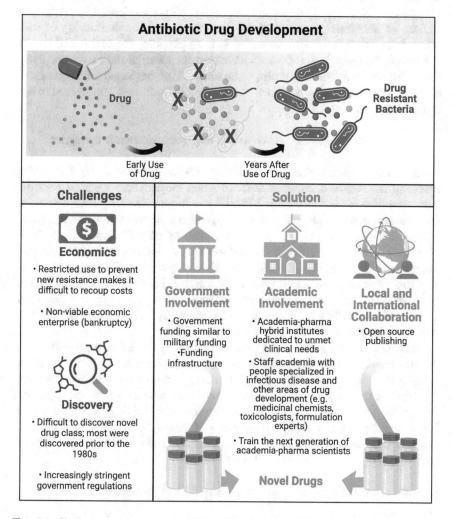

Fig. 6.1 Challenges and proposed solutions for antibiotic drug discovery. (Created with BioRender.com)

6.1.1 Reasons for the Lack of New Antibiotics

- Difficult to recoup cost of development

Antibiotics are no longer investment-grade products because of the current high-risk, high-return nature of the biotechnology-pharmaceutical industry. These generally acutely administered life-saving drugs are unable to price for their true value to patients, nor will society tolerate the exorbitant prices that many catastrophic rare diseases and diseases like cancer command. Moreover, as these new drugs are expected to specifically address resistant infectious organisms, their use is restricted to limited patient populations infected with resistant organisms. Even new broad-spectrum antibiotics will not see the vast market sizes experienced by earlier generations of antibiotics, because of the need for thoughtful stewardship of these agents to preserve their effectiveness as long as possible. Tropical infectious diseases also have severely constrained reimbursement potential and are unattractive to the high-return sensitive investment community. These neglected infectious agents also face future resistance liabilities and, with global warming, their spread to new geographic territories will increase.

- Antibiotic development is proven to be a non-viable economic enterprise

After some initial enthusiasm among the investment community for funding antibiotic development, excitement faded when the biopharmaceutical company Achaogen filed for bankruptcy in 2019, after reporting Zemdri (plazomicin injection—an aminoglycoside antibiotic with expanded coverage against several multi-drug resistant organisms) had net product sales of only $0.8 million for 2018. Melintra Therapeutics, which developed the fluoroquinolone antibiotic delafloxacin, also filed for bankruptcy in 2019, as did Aradigm, when FDA required a large new trial of its inhaled Cipro product. Tetraphase Pharmaceuticals, inventor of eravacycline (a different and improved tigecycline/tetracycline type antibiotic) with net sales of $4.3 million in six months, was acquired by La Jolla Pharmaceuticals in 2020 for a fraction of the original investment.

It seems wishful thinking that a long-term privately funded enterprise dedicated to critical antibiotic discovery and development can be sustained on an occasional public prize or bonus incentive system. Building biotech companies to create one-off antibiotics is exceedingly inefficient. These are one and done financially driven enterprises, not the long-term commitment required for building an evergreen pipeline of new antibiotics to replace those therapeutics that fall to resistance. Trying to shoehorn new antibiotic and parasitic drug invention and development into the current biotech/pharma business model is a nonstarter.

- Novel classes of antibiotics are particularly difficult to discover

Almost every class of therapeutically useful antibacterial drugs was initially identified before the early 1980s, and many came with liabilities that are unacceptable today. Even recent class entries have long histories. For instance, Linezolid received FDA approval in 2000, but the antibacterial activity of oxazolidinones was first noted in 1956. Daptomycin was discovered by Lilly in the late 1980s, abandoned after early-stage clinical trials revealed adverse effects on skeletal muscle, and revived by Cubist Pharmaceuticals when it acquired the rights and received FDA approval six years later in 2003. Retapamulin received regulatory approval in 2007, but the initial lead natural product was discovered in 1950. Fidaxomicin received FDA approval as a treatment for Clostridium difficile in 2011, decades after it was first discovered in the 1970s.

- The financial challenge

The demand for return on capital in the modern therapeutics industry has grown enormously over the past 60 years. This financial focus has significantly influenced the direction of new drug invention towards therapeutic markets that will maximize return on investment. In the absence of extraordinary pricing power or enormous market size, drug research and development (R&D) funding becomes uncertain or even unavailable for important therapeutic areas like bacterial and tropical infectious diseases. The impact of the right pricing power incentives will, in fact, create lucrative markets where none existed previously. This has been seen in biopharma for therapeutics addressing rare diseases, for cancer drugs, and for most injectable biological treatments.

It is highly unlikely that new antibiotics or other classes of drugs for neglected infectious diseases will be allowed the extraordinary pricing power of cancer or rare diseases. This results in an ongoing market failure to provide new and much needed medicines to replace the earlier generations of antimicrobials that lose efficacy against resistant organisms.

6.1.2   The Solution: Build, Fill, Expand, and Refill an Evergreen Anti-infectious Agent Pipeline

To address this shortfall now and into the future (because infectious organisms will continue to evolve resistance to each new generation of therapeutics), consideration should be given to establishing a network of applied medicines research and development institutes within a number of host universities. These institutes would focus exclusively on discovering and translating prototype anti-infectious agents into drug candidates to specifically address targeted resistant bacterial and parasitic diseases.

- A new funding mechanism

There are no consumer markets for tanks, missiles, fighter jets, bombers, or submarines. Yet, it is agreed that these must exist at great national expense to protect nations and their populations. To rectify this problem, governments have created an entire military funding and procurement infrastructure to design, engineer, and supply the military necessities of societies without which the inventors, designers, and suppliers would cease to exist. Governments fund military R&D and purchase the products at agreed upon prices and in agreed upon quantities. Much like the military R&D ecosystem, the infectious disease marketplace now requires a different and novel funding approach. One such model is proposed below.

- A new mindset

We first need to build a different reward/promotion/accountability mindset among the participants and leadership, as well as for the required interdisciplinary teams necessary to successfully advance product candidates to ensure a robust pipeline of drug candidates. It should be founded on an open-source approach with efficient and rapid dissemination of all results generated. The global community can then rapidly assess, provide input, supplement, participate in, and utilize the findings of any particular project under development in this enterprise.

- A new role for academia

New institutes should be formed within academia with the mission to create clinically testable drug development candidates addressing unmet bacterial and parasitic threats. These institutes are not typical small grant-funded endeavors, but dedicated academic-pharmaceutical industry hybrid institutes tasked and funded specifically to create and advance antimicrobial drug candidates for the most pressing resistant microorganisms. Rewards for the institute and staff would be based on the identification and advancement of infectious disease control agents through phase 2 "clinical proof of concept" trials. Publishing in open-source journals is important for rapid dissemination of new science, but not an end in itself. Situating these institutes within academia also allows for sharing of unique facilities and expertise and is infrastructure-efficient. This approach is reminiscent of the penicillin and antimalaria efforts during World War II.

- Staff academia with experts

These institutes will be staffed with personnel with expertise typically seen in an integrated pharmaceutical R&D operation. They would employ specialists in infectious disease biology, as well as microbiology and infectious disease pharmacology, medicinal chemistry, natural products and peptide chemistry, protein engineering, structural biology, fermentation/plant culture, drug formulation, informatics/artificial intelligence, project management, contracts and intellectual property management, Chemistry, Manufacturing, and Controls (CMC) outsourcing specialists, drug safety and toxicology specialists, regulatory specialists, biostatisticians, clinical trialists, medical writers, clinical operations managers, and infectious disease clinicians. However, much like a corporate drug R&D operation, scientific management will be responsible for program goals and progress, team oversight, timelines, resource allocation, and delivering product candidates. The management team will be responsible for keeping the pipeline full and candidates moving towards a prephase 3 handoff.

- Collaborate internationally

Since this is a confederation of institutes, not every facility would have to fully replicate all necessary functions. Resources could be shared among institutes, especially development-enabling resources like toxicology and fermentation facilities, clinical trialists, and biostatisticians, for example. The development functions would be added slowly as a pipeline of development candidates materialized. These university-based institutes should facilitate collaboration with specialists/experts in the local academic community. The chemical source material for these institutions can come from academic investigators around the globe, as well as internal drug discovery projects.

In fact, the supply could arise from external candidates stalled in their development, especially candidates sitting at the edge of the "valley of death" that can be commandeered to generate the first pipeline candidates. Work can be outsourced to academic scientists with special skills related to a particular infectious organism. Development activities, like clinical trials, toxicology studies, and CMC activities, can be outsourced to appropriate vendors and managed by experienced internal staff. Sufficient money must be allocated for ongoing support of these preclinical and clinical development activities. Development teams can use the same global contract development support infrastructure of contract research organizations (CROs) and contract manufacturing organizations (CMOs) that has evolved over the past 30 years to service the biotech and pharma industry.

Particular institutes might emphasize natural product mining and screening; address resistance by structural modification of older classes of antibiotics (a classic tried-and-true approach used by some biotech companies); molecular target-driven antibiotic drug invention; biological therapeutics such as antibodies, enzybiotics, and parasitic phages targeting specific bacterial infections; specific high-burden parasitic infection drug discovery; new target identification/validation coupled to compound screening efforts; biofilm disruption; quorum sensing disrupters; etc. with the goal of identifying potential therapeutic candidates for development. Interesting chemical matter from academic antibacterial and antiparasitic programs around the world can also be a source of development candidates if the discoverers are willing to engage in a collaborative development relationship. Many academic investigators with interesting antibacterial and antiparasitic compounds would now have a potential development-focused partner for their stalled product candidates. As stated above, the goal of this endeavor is an evergreen pipeline of anti-infectious agents meeting our current and future medical needs.

- Build academic knowhow

These institutes can also serve as training grounds for the next generation of applied bioscience scientists and clinicians interested in infectious disease control and drug development. As the last generation of practitioners with antibacterial drug discovery and development experience ages out, there is a need to also reestablish a pipeline of antibiotic drug developers. Without commencing practical training, this experience and expertise base might have to be totally rebuilt from scratch in the middle of a future crisis. The global expert base of infectious disease scientists and drug developers targeting parasitic diseases can also be enriched as those products are advanced into the clinic outside the US.

- Fund them like the army

Institute members can collaborate with their academic colleagues in infectious disease departments and/or collaborators and institutions around the world, with funding support for collaborations, as well. These institutes can award grants to experts outside the US who will expedite the advancement of programs. Such institutes can fund CMC and manufacturing work at CROs. This approach will require ongoing, new systems levels of funding, not the typical modest academic grant levels of funding. These institutes are the biological analog of the R&D weapons laboratories at Los Alamos and Lawrence Livermore without the security clearance requirement. This initiative should garner the same large long-term funding commitment for the everlasting war on infectious diseases. The oversight and strategy focus of these institutes will be directed and funded by a joint committee pulled from the Centers for Disease Control and Prevention (CDC) and National Institute for Allergy and Infectious Diseases (NIAID). The Department of Defense might also consider funding antimicrobial work targeting specific biothreats and infectious diseases that service members may encounter overseas.

These institutes will be tasked to deliver phase 3-ready anti-infectious disease therapeutic agents. At that stage, the oversight custodians and funding agencies of this work, NIAID and CDC, will take control of the product candidates and decide, in collaboration with FDA, on a roadmap for their continued development, approval, and a strategy for future drug manufacturing, procurement, distribution, and pricing. Any number of viable late-stage development, manufacturing, and supply possibilities can be envisioned.

6.1.3 We Can Address the Crisis Before It Is at Our Doors

The Need Is Now It is important to note many of today's key therapeutic interventions rely heavily on the availability of highly effective antibiotics. These include joint replacements, bone marrow and organ transplants, myelosuppressive oncology treatments, autoimmune disease treatments, bowel surgery, burn treatments, control of skin and stomach ulcers, use of indwelling ports and catheters, post-trauma surgeries, and even cosmetic procedures. As the efficacy of today's antibiotics wanes, many of these medical procedures will become too dangerous to patients. The time is now to address the infectious diseases problem before it becomes a greater crisis.

New agents for two exceedingly important unmet medical indications, resistant bacterial and parasitic infections, will continue to languish within the modest individual academic investigator community. Without a reimagination of the drug development ecosystem outside of the standard for-profit investment model, academia discoveries will just sit on the laboratory shelf. Without action now, tropical diseases will continue to plague a significant number of the world's population and patients worldwide will be at increased risk of death from previously treatable garden variety bacterial infections with mortality rates approaching those seen in the pre-antibiotic era. Pneumonia, tuberculosis, and diarrheal diseases could top the mortality tables again.

Should this approach prove viable in restocking the world's medicine cabinets with new infectious disease control agents, the model could be expanded to encompass other therapeutic arenas that are poorly served or abandoned by the current therapeutic creation industry. This could become especially important for critical areas that, in the future, will be far more constrained by limits placed on drug pricing power.

We cannot sit on our hands hoping someone, somehow will address this forever problem. We do not do that for our military preparedness concerns. Why should we default to an ostrich strategy for our eternal war against microbes? There are no magical solutions here. We must do something different, before we have a catastrophic problem on our hands.

The Bottom Line

Developing novel therapeutics for resistant microbes is a societal need that is not being addressed by the for-profit biopharmaceutical ecosystem. Much like the defense industry, federal governments should support research and product development in the war against drug resistant pathogens. We suggest establishing a collaborative group of government-funded research institutes dedicated to developing novel antimicrobial agents to support this critical need.

Key Terms and Abbreviations

Key Abbreviations

ADME	Absorption, Distribution, Metabolism, and Excretion
CDC	Centers for Disease Control and Prevention
CMC	Chemistry, Manufacturing, and Controls
CMO	Contract Manufacturing Organization
CRO	Contract Research Organization
NIAID	National Institute for Allergy and Infectious Diseases
R&D	Research and Development
FDA	US Food and Drug Administration

Reference

Murray CJ, Ikuta KS, Sharara F et al (2022) Global burden of bacterial antimicrobial resistance in 2019: a systematic analysis. Lancet 399:629–655

Correction to: Therapeutics and Diagnostics Discovery

Daria Mochly-Rosen, Kevin Grimes, Rami N. Hannoush, Bruce Koch, Gretchen Ehrenkaufer, Daniel A. Erlanson, Julie Saiki, Jennifer L. Wilson, Shelley Force Aldred, Adriana A. Garcia, Jin Billy Li, Rosa Bacchetta, Maria Grazia Roncarolo, Alma-Martina Cepika, Harry Greenberg, Steven N. Goodman, and Michael A. Kohn

Correction to:
Chapter 2 in: D. Mochly-Rosen, K. Grimes (eds.), *A Practical Guide to Drug Development in Academia*,
https://doi.org/10.1007/978-3-031-34724-5_2

The original version of Chapter 2, "**Therapeutics and Diagnostics Discovery**," was previously published without **Dr. Gretchen Ehrenkaufer** being credited as the chapter author. The book has been updated with these changes.

The updated version of this chapter can be found at
https://doi.org/10.1007/978-3-031-34724-5_2

Author Biographies

Shelley Force Aldred, PhD, is a serial entrepreneur and experienced drug developer. Dr. Force Aldred is CEO and co-founder of Rondo Therapeutics, a biopharmaceutical company developing bispecific therapeutic antibodies for oncology indications. Previously, she served as VP for Preclinical Development at Teneobio Inc., a highly successful multispecific therapeutic antibody company. In the past, Dr. Force Aldred was COO and co-founder of SwitchGear Genomics, a venture-backed functional genomics platform company. Prior to founding SwitchGear Genomics, Dr. Force Aldred was a Scientific Director on Stanford's ENCODE Project and received her PhD from Stanford University.

Rosa Bacchetta, MD, is an Associate Professor of Pediatrics at Stanford University and has long-standing experience investigating the mechanisms of immunological tolerance, specifically in pediatric patients after hematopoietic stem cell transplantation, and those with genetic diseases of the immune system. She completed her pediatrics residency at the University of Turin, then received training in molecular and cellular immunology at the DNAX Research Institute of Molecular and Cellular Biology. She then worked for fifteen years at the San Raffaele Scientific Institute (HSR-TIGET), where she focused on dissecting the genetic and immunological basis of primary immune-regulatory diseases that might be treated by gene therapy. Dr. Bacchetta's research focuses on dissecting the role of FOXP3 and Treg cells in human immune responses, with a goal of establishing cell and gene transfer-based therapies for IPEX Syndrome. She is sponsoring a phase 1 gene therapy trial using autologous Treg-like cells to treat IPEX patients.

Mark Backer, PhD, received a BS in Chemistry from Stanford University and a PhD in Chemical Engineering from the University of Washington. He has worked to develop biopharmaceuticals since joining Genentech as its seventh employee in 1978 and has participated in the development of seven commercial products. He is currently General Manager of Alava Biopharm Partners, a consulting group focused

D. Mochly-Rosen, K. Grimes (eds.), *A Practical Guide to Drug Development in Academia*, https://doi.org/10.1007/978-3-031-34724-5

on CMC and regulatory support for product developers. He also supports the SPARK program in an advisory role.

Glenn Begley, MBBS (MD-equivalent), PhD, is an oncologist and hematopathologist whose academic and commercial roles include board level and senior positions in Australia, the US and the UK. He currently serves as Co-Founder and Head of Drug Discovery at Parthenon Therapeutics and as Senior Scientific Consultant to Certara and to BridGene Biosciences. He was CEO of BioCurate, an Australia-based joint venture between the University of Melbourne and Monash University. Previously, he was Chief Scientific Officer at Akriveia Therapeutics, Global Head of Hematology and Oncology Research at Amgen, Senior Vice President and CSO at TetraLogic Pharmaceuticals, and executive director of the Western Australian Institute of Medical Research. His early research first described human G-CSF, and in later clinical studies he first pioneered the use of G-CSF-"mobilized" blood stem cells (so-called "stem cell transplantation"). He received his PhD from the University of Melbourne and was elected Fellow of the Australian Academy of Health and Medical Sciences.

Rebecca Begley, PhD, is an Executive Director of Project Management at Terns Pharmaceuticals. Prior to that, she worked at Gilead Sciences, Inc., as a Clinical Pharmacologist and Regulatory Project Manager, and at KAI Pharmaceuticals as a Scientist and Project Manager. She received her undergraduate degree from Barnard College and her PhD from Stanford University in 2004.

Terrence F. Blaschke, MD, is Emeritus Professor of Medicine and Molecular Pharmacology at Stanford University School of Medicine. His early research included the study of drugs used in HIV-infected patients, which lead to a particular interest in the access and adherence to these drugs in low- and middle-income countries. At Stanford University, he served roles including Associate Dean for Medical Student Advising and Associate Director, Stanford General Clinical Research Center. After becoming emeritus, he joined the Bill and Melinda Gates Foundation as a Senior Program Officer, Global Health Discovery and Translational Sciences, and has published extensively on medication adherence.

Robert F. Booth, PhD, has more than 30 years of experience in drug discovery and development in both large and small biopharmaceutical companies, both in the US and in Europe. He also has experience in business development, venture capital, and in academia. In the recent past, he has served as co-founder and Chairman of the boards of Curasen Therapeutics (a SPARK project) and Ab Initio Biotherapeutics and more recently as Chairman of Myoforte Inc. (another SPARK project). Prior to those roles, he was founder and CEO of Virobay Inc., Operating Partner at TPG Biotech, Chief Scientific Officer at Celera Genomics, and Senior Vice President at

Roche. He currently serves as a board member and on the scientific advisory boards of several biotech companies. He received his BSc and PhD in biochemistry from the University of London.

Peter Boyd, JD, MBA, is a healthcare entrepreneur with a background in business, science, and law. He was a co-founding member of Harpoon Medical where he negotiated and managed all aspects of a complex structured financing and the subsequent acquisition by Edwards Lifesciences. He worked as an associate until 2012 in the Silicon Valley office of Latham & Watkins LLP, where he represented startup companies and venture capitalists. Mr. Boyd earned his BS in Biology from the University of North Carolina at Chapel Hill and a joint JD–MBA degree from the University of Virginia.

Jennifer Swanton Brown, RN, CCRP is Director of Clinical Research Quality and Assistant Dean for Compliance, Regulatory and Quality in the office of the Senior Associate Dean for Research, Stanford School of Medicine.

Alma-Martina Cepika, MD, PhD, is a physician scientist with an extensive background in human immunology and genomics. She is an Instructor of Pediatrics at Stanford University School of Medicine and a current recipient of several awards in the field of cell and gene therapies, including the SPARK Translational Research Grant, the National Blood Foundation Early Career Research Grant, and the American Society for Gene + Cell Therapy Career Development Award. She completed her postdoctoral training at the Baylor Institute for Immunology Research and obtained her MD degree at the University of Zagreb School of Medicine in Croatia. There, she also completed an international master's course in Leadership and Management of Health Services, founded in collaboration with the London School of Economics and Political Sciences. She received an early exposure to biopharma during her PhD at Croatia's Institute of Immunology, which was a leading developer of vaccines, anti-venoms, and other medicinal products for southeast Europe.

Leon Chen, PhD, MBA, is a Partner at The Column Group, focused on investing and new company formation. Prior to joining TCG, he was a venture partner at OrbiMed Advisors and a Partner at Skyline Ventures. He was a cofounder of KAI Pharmaceuticals and has held founder or early startup roles in multiple biotech companies including Eikon Therapeutics, Escape Bio, Adicet Bio, and TranscripTx. As a biotech entrepreneur, he has experience in drug discovery and building a research organization and has taken drugs from the laboratory into clinical development. Dr. Chen has a PhD in Molecular Pharmacology from Stanford University School of Medicine and an MBA from the Stanford Graduate School of Business.

Eugenio L. de Hostos, PhD, MBA, is Senior Director at the Global Health portfolio at Calibr, the drug discovery and development division of Scripps Research. Previously, Dr. de Hostos was Director in the Drug Development Program at PATH and The Institute for OneWorld Health. He received his BS from Yale University, his MBA from Presidio Graduate School, and his PhD from Stanford University.

Emily Egeler, PhD, is Director of Regulatory Operations and Procedures for the Center for Cancer Cell Therapy at Stanford University School of Medicine, where she supports investigator-initiated studies with novel cell therapies. Her previous roles include Regulatory Specialist with the Cancer Clinical Trials Office and Program Coordinator for the SPARK Translational Research Program, both at Stanford University School of Medicine. She received her PhD in Chemical and Systems Biology from Stanford University.

Gretchen Ehrenkaufer, PhD, did her doctoral work at University of California, San Francisco in the lab of Dr. Thomas Kornberg, where she studied Hedgehog signaling in Drosophila development. After graduating with a PhD in cell biology, she began infectious disease work in Dr. Upi Singh's lab at Stanford University, studying the protozoan parasite Entamoeba histolytica that causes amoebic dysentery. While in Dr. Singh's lab, she worked on many diverse projects, including investigating the transcriptional changes and molecular mechanisms of amebic development, and analysis of small RNA populations. Her involvement with SPARK began in 2016 as part of a project to develop new therapeutics to treat amoebic infection. This work has resulted in multiple publications and identified promising drug candidates, including several with pan-parasitic activity and others that are FDA approved or in late-clinical stages. She has been a Program Manager at SPARK since 2021.

Daniel A. Erlanson, PhD, is VP of Innovation and Discovery at Frontier Medicines. Before that, he co-founded Carmot Therapeutics, whose Chemotype Evolution technology he co-developed. Dr. Erlanson received his BA in Chemistry at Carleton College and his PhD in Chemistry at Harvard University in the laboratory of Gregory L. Verdine. He was an NIH postdoctoral fellow with James A. Wells at Genentech. From there, he joined Sunesis Pharmaceuticals at its inception, and went on to develop fragment-based technologies and to lead medicinal chemistry programs. As well as co-editing two books on fragment-based drug discovery, Dr. Erlanson is editor of Practical Fragments (http://practicalfragments.blogspot.com/).

Lyn Frumkin, MD, PhD, previously worked at biotechnology companies Amgen, Inc. and ICOS Corp., where he provided broad expertise, including leading clinical teams conducting global programs for the successful approval of pivotal programs in North America, Europe, and Japan. Dr. Frumkin was a Consulting Professor at Stanford University School of Medicine and has been an advisor to SPARK since 2010. He is also a Fulbright awardee, which allows service as an expert consultant on programs related to transitional medicine/global health at foreign academic institutions. Dr. Frumkin received his MD-PhD from the University of Washington with postgraduate training in internal medicine at Stanford University and subsequent subspecialty training in Neurology and Infectious Diseases/Virology.

Nicholas Gaich is the founder and CEO of Nick Gaich and Associates, a firm dedicated to providing executive coaching, leadership development, strategic planning, and operational performance. He has more than 46 years of experience with expertise ranging from executive coaching, organizational development, change management, supply chain management, customer service, hospital service line economics, clinical research administration, business development, and marketing. Mr. Gaich also serves as the CEO/President of the Morgan Hill Chamber of Commerce. Mr. Gaich retired in 2012 as Assistant Dean of Clinical and Translational Research Operations, Stanford Center for Clinical and Translational Research and Education at Stanford University School of Medicine. He also held a two-year appointment on the Consortium Management Group at the National Center for Research Resources, National Institutes of Health. He received his BBA with an emphasis in Healthcare Administration from National University of San Diego.

Adriana A. Garcia, PhD, is an Academic Professional at SPARK at Stanford. Prior to that, she received her PhD in Chemical and Systems Biology from Stanford University and a BS in Biochemistry from San Francisco State University. Her doctoral work investigated the stabilization of protein oligomers as a therapeutic strategy for a common enzymopathy known to cause hemolytic anemia.

Steven N. Goodman, MD, MHS, PhD, is Associate Dean for Clinical and Translational Research and Professor of Epidemiology and of Medicine at Stanford. He directs the Stanford Program on Research Rigor and Reproducibility (SPORR), whose aim is to increase the adoption of best scientific practices throughout the Stanford School of Medicine. He leads a variety of training programs in epidemiology and clinical research, and his research is in scientific and statistical inference, with connections to research, policy, and ethics. Outside of Stanford, Dr. Goodman serves as chair of the PCORI Methodology Committee and is senior statistical editor of the Annals of Internal Medicine and scientific advisor to the national Blue Cross-Blue Shield technology assessment program. He was awarded the Spinoza Chair in Medicine from the University of Amsterdam for his work in inference, the Lilienfeld award from the American College of Epidemiology for his lifetime contributions to the field of epidemiology, and in 2020 was elected to the National Academy of Medicine.

Julie Papanek Grant, MBA, is a venture capitalist and pharmaceutical executive. She has incubated, founded, and invested in a number of new pharmaceutical companies as a General Partner at Canaan. Prior to joining Canaan, she worked across Development and Commercial functions at Genentech and subsequently Roche. After graduating from Yale University with a BS in Molecular Biophysics and Biochemistry, she received an MPhil in BioScience Enterprise from Cambridge University and an MBA from the Stanford Graduate School of Business.

Harry Greenberg, MD, is the Joseph D. Grant Professor of Medicine and Microbiology and Immunology and an Associate Dean for Research at Stanford University School of Medicine. He has been an active NIH-funded investigator for almost 40 years during which time his studies have focused primarily on viruses that infect the GI tract, liver, or respiratory tree. Dr. Greenberg was the Chief Scientific Officer at biotechnology company Aviron (now MedImmune Vaccines), where he played an important role in bringing the live attenuated influenza vaccine to licensure. He was also a long-term advisor to Bharat, an Indian Vaccine company that has developed novel vaccines for low-income countries such as a new rotavirus vaccine and a novel and highly effective Salmonella Typhi conjugate vaccine.

Kevin Grimes, MD, MBA, is a Professor of Chemical and Systems Biology at Stanford University and Co-Director of SPARK at Stanford. He received his MD from Brown University. Dr. Grimes began his career as a Clinical Assistant Professor of Medicine at Stanford, where his primary duties included the teaching and practice of internal medicine. Dr. Grimes received a Hartford Foundation Fellowship to study health economics and obtained an MBA at the Stanford Graduate School of Business. He was subsequently selected as a White House Fellow and assigned to the Department of Defense, where he served as Special Assistant to the Secretary. He spent fifteen years in industry, working in the medical device, life science consulting, and biotechnology sectors prior to returning to Stanford to co-direct SPARK. Dr. Grimes also teaches graduate student courses on drug discovery and development and continues to teach and practice internal medicine. He has received the David Rytand Award for Excellence in Clinical Teaching and the Faculty Award for Excellence in Graduate Teaching.

Grace Hancock, PhD, is Program Manager for the Global Health portfolio at Calibr, the drug discovery and development division of Scripps Research. She received her BS from the Johns Hopkins University and her PhD from University of California, Los Angeles.

Rami N. Hannoush, PhD, spent fifteen years at Genentech where he was a Senior Group Leader, and he led both early- and late-stage discovery biology programs in Ophthalmology, Immunology, and Oncology, with a focus on drug discovery and development, as well as target discovery and validation. His laboratory in translational biology at Genentech focused on new drug modalities and understanding the role of protein–protein interactions in stem cell signaling pathways for cellular reprogramming and tissue regeneration, and his group made important contributions in the fields of Wnt signaling and protein lipidation. He obtained his PhD in Chemistry from McGill University and did his postdoctoral training in chemical biology and cell biology at Harvard University. He has received numerous academic and professional awards, authored more than 60 peer-reviewed scientific publications, held several patents, and served on multiple scientific editorial boards as well as advisory boards for organizations spanning industry and academia. He is also currently an adjunct Professor of Chemical and Systems Biology at Stanford University.

Karin Immergluck, PhD, is the Executive Director of the Stanford Office of Technology Licensing and Industrial Contracts Office. Prior to joining Stanford, she was the Executive Director of the Office of Technology Management at University of California, San Francisco. She obtained her PhD in Molecular Developmental Genetics from the University of Zurich.

Carol D. Karp is the Chief Regulatory Officer for Prothena Biosciences. She previously held leadership roles with Johnson & Johnson, CV Therapeutics, Esperion, and VIVUS. Her experience in the biotechnology/pharmaceutical sector spans the global development and commercialization of products to address unmet needs for therapeutic areas including neurodegenerative, cardiovascular, renal, metabolic, immunologic, and addiction diseases and disorders. Ms. Karp has been a SPARK advisor since 2009. She earned her BA in Biology from the University of Rochester, where she currently serves as Vice Chair of the Board of Trustees.

J. Jekkie Kim, JD, MD, MBA, LLM is a Partner at Latham & Watkins LLP. Dr. Kim advises healthcare, life sciences, and technology companies and their investors on a range of intellectual property transactional matters. She draws on her background as a medical doctor and has developed a particular focus on digital health, agricultural technology, and cross-border transactions that involve markets such as China. She received an MD from College of Medicine, Yonsei University, JD from Boston University, MBA from Case Western Reserve University, and LLM from Yonsei University Department of Medical Law and Ethics.

Nina Kjellson is a tenured venture capitalist with over 25 years of biotechnology and digital health investing experience. She is a General Partner at Canaan Partners and was previously a General Partner at InterWest Partners. Prior to InterWest, she was an Associate at Bay City Capital, a life sciences merchant bank and an Analyst at Oracle Partners, a healthcare hedge fund. She began her career conducting research for the Kaiser Family Foundation. She is a Health Innovation Fellow at the

Aspen Institute and a member of the Leadership Council of the Health Innovation Center at Oliver Wyman. She is a member of the boards of directors of Girle Effect, Essential Access Health and Life Science Cares' Bay Area chapter. She has a BA in Human Biology from Stanford University.

Bruce Koch, PhD, received his PhD in Cell and Developmental Biology from Harvard University and completed his postdoctoral studies at University of California, Berkeley. He joined the discovery research group at Syntex rising to the position of Director of Discovery Technologies at Roche Pharmaceuticals. He is currently the Director of the Stanford's ChEM-H/Chemical and Systems Biology High-Throughput Screening Knowledge Center at Stanford University.

Michael A. Kohn is Professor Emeritus of Epidemiology and Biostatistics at University of California, San Francisco and was an attending emergency physician at Mills-Peninsula Medical Center in Burlingame until May 2018. He teaches in UCSF's Training in Clinical Research (TICR) Program on research methods, clinical epidemiology, and database management. He also provides consultations to clinical investigators who need help with study design, sample size calculations, data management, and statistical analysis. His primary research interest is diagnostic testing. He co-authored the clinical epidemiology textbook *Evidence-Based Diagnosis*, is a contributing author to Hulley et al., "Designing Clinical Research," and created the UCSF Clinical & Translational Science Institute (CTSI) online sample size calculators (www.sample-size.net). Dr. Kohn is also Chief Medical Officer of QuesGen Inc. He received an AB from Stanford University, an MPP from Harvard Kennedy School, and an MD from Stanford and trained in emergency medicine (board certified) at the Denver Affiliated Program.

Jin Billy Li, PhD, is an Associate Professor of Genetics at Stanford University. He received his bachelor's and master's degrees at Tsinghua University and his PhD degree from Washington University in St. Louis. After his postdoctoral training at Harvard Medical School, he started his laboratory at Stanford in 2010 where he has focused on studying RNA editing mediated by ADAR enzymes. His laboratory focuses on two aspects of ADAR RNA editing. One is the biological function of RNA editing to evade dsRNA-mediated autoimmunity, which has led to new approaches to treating cancer and inflammatory diseases. The other is to harness the ADAR enzyme for site-directed RNA base editing that overcomes the challenges of CRISPR/Cas-based DNA editing and holds great potential for treating rare and common diseases.

Robert Lum, PhD, is a veteran in drug development with experience in both large and small pharmaceutical companies. He has served in a variety of positions, including Senior Director, Pharmaceutical Operations at Concentric Analgesics; Senior Director, Technical Operations at Aquinox; Director of Process Development at

Pfizer (formerly Medivation); Executive Director, Process Development and Manufacturing at Geron Corp.; Vice President, Technology and Preclinical Development at Telik, Inc.; Assistant Director of Medicinal Chemistry at CV Therapeutics; Senior Scientist at Arris Pharmaceutical; and Assistant Senior Investigator at SmithKline Beecham. He received his BS in Chemistry from University of California, Berkeley, and his PhD in Organic Chemistry from Massachusetts Institute of Technology.

Collen Masimirembwa, PhD, is the founder, President, and CEO of the African Institute of Biomedical Science and Technology (AiBST). He has over 25 years of experience in pharmacokinetics and pharmacogenomics for drug discovery and development, ten of them as a Principal Scientist at AstraZeneca Pharmaceutical company. He is President of the African Pharmacogenomics Consortium and chairs the Global Pharmacogenomics Research Network. His pioneering work in pharmacogenomics in Africa has won him many awards including the Human Genome Organisation Award and the Bill and Melinda Gates Foundation Calestous Juma Science Leadership Award to expand pharmacogenomic research across Africa. He is also the Director of SPARK GLOBAL Africa.

Ted McCluskey, MD, PhD, has over 25 years of experience in biotechnology, pharmaceutical, and diagnostic development designing, implementing, and analyzing clinical trials. He has worked in multiple disease areas and participated in four product approvals. Past positions include Chief Medical Officer and VP Clinical at AVIIR, Inc., Senior Director of Clinical Research at Johnson & Johnson/Scios Inc., and Senior Medical Director at Genentech/Roche. He is an interventional cardiologist and was previously on the faculty at University of Cincinnati, trained at the Cleveland Clinic and University of California, San Francisco, and received his MD and PhD from Washington University in St. Louis. Dr. McCluskey was also formerly active in Sand Hill Angels, having served as its President and on the Board of Directors. He has been an advisor at SPARK since 2011.

Dirk Mendel, PhD, is a veteran in drug development with close to 30 years of experience in large and small biotech and biopharma companies. He has worked in a variety of therapeutic areas, most notably virology and oncology, and has contributed significantly to the development of therapeutic small molecules, antibodies, and peptides, including 7 that have made it to market. He has worked in all phases of drug development, from early research through clinical proof-of-concept and regulatory approval, and has spent the past 20 years focusing on the translational aspects of drug development, leading translational efforts from lead optimization through clinical proof-of-concept to understand patient identification and stratification, and identifying and understanding the relationships between compound exposure and activity/toxicity to enable clinical candidate selection and facilitate early clinical development. Dr. Mendel received his BS in Engineering from Stanford University and his PhD in Physiology from Dartmouth Medical School.

Alan Mendelson, JD [died October 8, 2021], was a founding partner at Latham & Watkins LLP's Silicon Valley office and cochaired the firm's Emerging Companies Practice and Life Sciences Industry Groups. He served as a member of the University of California Board of Regents and the Boards of Trustees of the UC Berkeley Foundation and The Scripps Research Institute, served on the board of the California Life Sciences Association, and was the corporate secretary for many public and private companies. Throughout his career, Mr. Mendelson mentored many life sciences-focused attorneys and biotech executives. He received his AB from University of California, Berkeley, and his JD from Harvard University.

Daria Mochly-Rosen, PhD, is a Professor of Chemical and Systems Biology and the George D Smith Professor of Translational Medicine at Stanford University School of Medicine. She is Founder and Co-Director of SPARK at Stanford and the President of SPARK GLOBAL. She leads a multi-disciplinary research lab and has developed a number of drug leads for human diseases with a particular interest in mitochondrial biology and pathology. In 2003, her lab's basic research led to the founding of KAI Pharmaceuticals, where she served as CSO for one year and as chair of the Scientific Advisory Committee and a member of the Board of Directors after her return to academia. KAI was subsequently acquired by Amgen and KAI's drug, Parsabiv, has been approved. She also co-founded ALDEA Pharmaceuticals (now licensed to Foresee) and Mitofina Therapeutics. Dr. Mochly-Rosen previously served as chair of her department and the Senior Associate Dean for Research and continues to teach graduate-level classes at Stanford University School of Medicine. Her leadership in translational research efforts in academia led to awards from Accelerating Australia (2019), Cures Within Reach (2020), Xconomy Award (2020), and California Life Sciences Pantheon award (2022). Dr. Mochly-Rosen received her PhD in Chemical Immunology from the Weizmann Institute of Science and was a postdoctoral biochemistry fellow at University of California, Berkeley.

Judy Mohr, PhD, JD, received her PhD in Chemical Engineering from the University of Texas at Austin and her JD from Santa Clara University. She has extensive experience working in patent law with an emphasis in pharmaceuticals. Dr. Mohr is a partner at McDermott Will & Emery LLP (Silicon Valley Office). She also supports the SPARK program in an advisory role.

Christopher M. Reilly, JD, is Senior Director, Legal at Lyft Inc. He previously served as an associate at the Silicon Valley office of Latham & Watkins LLP and received his JD from the University of Virginia School of Law.

Maria Grazia Roncarolo, MD, is the George D. Smith Professor in Stem Cell and Regenerative Medicine, Professor of Pediatrics and of Medicine, director of the Center for Definitive and Curative Medicine, and co-director of the Institute for

Stem Cell Biology and Regenerative Medicine at Stanford University. A pediatric immunologist by training, she earned her medical degree at the University of Turin. She worked at DNAX Research Institute for Molecular and Cellular Biology, where she contributed to the discovery of novel cytokines. As director of the Telethon Institute for Gene Therapy at the San Raffaele Scientific Institute in Milan, Dr. Roncarolo developed novel approaches in cell and gene therapy. Her work led to the discovery of *ex vivo* gene therapies for genetic diseases of the immune system, including ADA-SCID and WASP, and metabolic diseases such as metachromatic leukodystrophy. The landmark stem cell gene therapy treatment for ADA-SCID was the world's first to be approved by the European Medicines Agency (EMA) under the brand name Strimvelis® in May 2016. She discovered a new class of T cells and is leading clinical trials using cell therapeutics to prevent immune mediated diseases. She is a co-founder of Graphite Bio, which is developing a new class of therapies to correct genetic defects in people with serious and life-threatening diseases. Dr. Roncarolo established the Stanford Center for Definitive and Curative Medicine to cure patients with currently incurable diseases through the development of innovative stem cell and gene-based therapies.

Werner Rubas, PhD, has 30 years of biotech and pharmaceutical industry experience and is currently VP of Preclinical Development at Sutro Biopharma. Dr. Rubas was previously at Nektar Therapeutics, most recently as Executive Director in Non-Clinical Pharmacokinetics and Pharmacodynamics. Prior to Nektar, Dr. Werner was Associate Director of the Drug Metabolism and Pharmacokinetics group at Roche. He has been a SPARK advisor at Stanford University since 2010 and teaches classes on drug development at UC Berkeley Extension. Dr. Rubas earned his PhD from ETH Zurich and received his pharmacy license from the School of Pharmacy at ETH Zurich.

Julie Saiki, PhD, is co-founder and Chief Operating Officer of Azora Therapeutics, a clinical-stage biotechnology company developing novel small molecules for the treatment of inflammatory diseases with high unmet needs. Previously, she was a management consultant at McKinsey & Company, where she worked in the pharmaceutical and medical products practice advising big pharma, biotech, and healthcare companies on strategy and operations. Dr. Saiki has an MS in Medicine and PhD in Chemical and Systems Biology from Stanford University and is a former Fulbright recipient.

Steve Schow, PhD, is an Adjunct Professor in the Chemical and Systems Biology Department at the Stanford University School of Medicine and has been a SPARK Advisor since 2009. Dr. Schow was the Vice President of Research and Development at Telik, Inc. until his retirement in 2014. He has more than 40 years of pharmaceutical, biotech, and agrichemical industrial R&D experience. His work in drug R&D spans a wide range of medical indications, as well research on pest control agents. Dr. Schow holds a doctorate in chemistry from University of California, San Diego, and completed postdoctoral training at University of California, Los Angeles, and the University of Pennsylvania.

Michael Taylor, PhD, is the Founder and Principal at Non-Clinical Safety Assessment, a consulting firm specializing in the development of drugs and medical devices. Dr. Taylor has more than 20 years of R&D experience in the pharmaceutical industry and has served on executive teams. He is a board-certified toxicologist and holds PhD and MS degrees in Toxicology from Utah State University with postdoctoral training at the NIH and CNRS of France. He has been a SPARK advisor for the past 6 years.

John Walker is a biotechnology veteran with over 40 years of experience in the healthcare and biopharmaceutical industries. He received a BA in History from the State University of New York at Buffalo and is a graduate of the Advanced Executive Program, J.L. Kellogg Graduate School of Business at Northwestern University. Mr. Walker has served on the Board of Directors of over three dozen life sciences companies. He served on the Board of Lucille Packard Children's Hospital and is currently a Trustee at the University of Puget Sound and the Board of Packard Children's Health Alliance.

Mary Wang, PhD, is Senior Director of Strategic Planning at Scripps Research. She received her BS from Massachusetts Institute of Technology and her PhD from Northwestern University.

Jennifer L. Wilson, PhD, is an Assistant Professor in Bioengineering at University of California, Los Angeles. Her research aims to use protein–protein interaction models to anticipate drug effects and identify new targets for untreated diseases. She has pursued training at FDA, within biotech (at Merrimack Pharmaceuticals, Genentech) and with SPARK to increase the translational potential of her network models. She received her BS in Biomedical Engineering from the University of Virginia and her PhD in Biological Engineering from Massachusetts Institute of Technology.

Haim Zaltzman, JD, is a Partner at Latham & Watkins LLP and chairs its San Francisco Bay Area Finance Practice. He focuses primarily on healthcare and technology-related private equity, growth equity, and emerging growth financing transactions. Mr. Zaltzman has also been featured on CNBC, Bloomberg, The Washington Post, CFO Magazine, The Recorder, Law360.com, The Daily Journal, the International Financial Law Review, and VCExperts.com for his debt expertise. He received his BA from Stanford University, MA from Russian Academy of Sciences, and JD from Harvard Law School.

Index